现代主义：东西方文化的杂合

Modernism: A Hybrid of Eastern and Western Culture

现代主义：东西方文化的杂合

［韩］俞炫准　著

中译　［韩］太贞姬　江岱

英译　［韩］崔胤雅　［韩］俞炫准

Modernism:

A Hybrid of Eastern and Western Culture

Hyunjoon Yoo

Chinese Translation
Junghee Tai
Jiang Dai

English Translation
Yoonah Choi
Hyunjoon Yoo

同济大学出版社
TONGJI UNIVERSITY PRESS

序：东方、西方与自我

三年多以前，我第一次看到这本书的韩文原版，作者以东方文化视点对现代主义所作的新诠释当时给我留下很深的印象。也很喜欢这本书的文风，作者用大量历史和文化方面的案例与比较阐明自己的反思，观点鲜明，简明易读，能够体会到作者所具有的东西方教育背景和文化底蕴，又完全没有通常理论著作的繁冗，让人觉得这本书的风格正是作者所青睐的现代主义。俞炫准教授是开业建筑师，又在大学承担重要教职，是韩国新生代建筑师的典型人物，他思考的现代主义根源的问题代表了韩国年轻一代建筑师对今后建筑发展方向的一种深层次思考。

关于“东方与西方”，当下，我们身边正在发生很大变化——在中国的城市与建筑领域，我们与欧美国家的交流与合作以超乎想象的速度日益频繁和普及。以同济大学建筑与城市规划学院为代表，学生有越来越多的机会到国外大学留学，除了以往学生个人申请的情况外，国内外合作的双学位硕士课程大批涌现出来。日益频繁的交往过程中有一个明显的变化，欧美国家以往的强势权威性减弱，这种减弱从经济政治领域传递到专业领域，大量欧美建筑和城市规划院校转向中国等正处于城市化发展阶段的国家寻找题目和学科发展空间，寻求对话，与大批中国学生去西方接受再启蒙教育形成逆向。由此，重新审视东西方关系，重新认识和定位自我是当下中国建筑教育和专业实践面临的严肃问题。

过去三年里，我有相当长时间在新加坡国立大学做访问教授，教城市设计，做有关亚洲城市的研究和考察，并不断往返于上海和新加坡，兼顾处于不同状态下的两方面工作。这段特殊经历让我看到了与东西方交流同等重要的东方与东方的关系，亚洲内部交流的意义是东西方交流无法替代的。从功利角度而言，今天中国城市化进程中许多具体问题在先行的亚洲城市中各有其解决办法，分析这些经验比学习西方更有效；更重要的是，亚洲视野下的交流提供了一个新视点理解自我——可以从具有相似文化背景和发展轨迹的近邻身上看到自

Foreword - East, West and Self

Three years ago I met with the original Korean edition of this book and it impressed me of the interpretation of Modernism from an East-West comparative perspective. I like the style of the book – it actually tackles kind of metaphysical issue, as author mentioned, the hereditary genealogy of modernism architecture, but not in a way of talking theoretically. Plenty of historic and cultural cases from both East and West are elaborated, which identify author's strong knowledge background from both worlds and make the book interesting, clear and easily-read. Hyunjoon Yoo runs practice actively and holds important position at university, his re-thinking on modernism partly shown in this book reflects a vision of young generation architects in South Korea about architecture in this new century.

We are in a changing time in terms of East and West. In China in fields of architecture and urbanism, links with West has been increasing at an incredible rate in the past three years. My school, the College of Architecture and Urban Planning, Tongji University, have launched a dozens of double-degree master programs with European and American architectural schools, which makes most of gradate students have opportunities to study abroad. Meanwhile, a lot of schools from West have come to Asian, especially to China seeking for research topics and opportunities for developing disciplines. Under such exchange situation, re-considering East-West link, re-evaluating and redefining "self" is serious for teaching and practicing architecture and urbanism in China.

Learning each other among Asian countries and cities is of the same importance as East-West links. I had a special experience in past two years, teaching urban design in the National University of Singapore as Visiting Associate Professor, traveling a lot of Asian places for Asian city study, meanwhile keeping my works at Shanghai moving on. The frequent shifting inside, outside China, and among different situations in Asia, has revealed me the significance of learning from Asia for China. There are respective solutions in some developed Asian cities for many concrete issues China cities confronting currently. More important, it provides a view point for self-reflecting by comparison with neighbors, developed or developing, with comparable cultural and social background and similar developing trajectories. This view point can only be reached by East-East approach. Modernism discussed by Hyunjoon Yoo in this book, and other discourse closely related such as modernity, perhaps are not key concerns in West

己的影子，从而更深刻地认识自我。这一点在东西方关系中很难体会到，二者的差异和差距太大。俞炫准教授用东方文化视点对现代主义进行反思和溯源，现代主义及现代性等问题，在西方已经不是问题，但恰恰是亚洲城市与建筑发展绕不过去的思想“门槛”。

此外，当我看到一个并非全貌的亚洲，尤其是看到一些并未国际化，将来也未必完全国际化的亚洲小城镇之后，第一反应是想知道中国大量的普通中小城市目前的状态，但知之甚少，这方面的资料也十分贫乏。我们太忙，日以继夜地为这些城市建设提供图纸，以至于没有时间去了解这些城市。许多国家和城市都经历过快速城市化过程，一战后欧洲大量住宅建设促进了现代主义建筑思想的普及，二战后更大规模的城市建设推广了现代主义城市规划思想。尽管西方在进入后现代时期后对二者大加批评，但就亚洲情况而言，现代主义与当时社会需求之间的合理契合仍然是值得学习的进步思想。当下的城市化进程为城市与建筑学科发展提供了一个历史契机，我认为，仔细理解自我，研究西方对应历史阶段的发展规律，提出适合于中国实际情况的新思路，不盲目跟随今天的西方潮流，是实现东方与西方平等对话的出路。

俞炫准教授这本书是同济大学出版社新推出的“倒影”丛书的第一本，丛书的主题是探讨现象和形式背后的思想“种子”。我相信本书所探讨的现代主义、东方和西方话题必然引发中国学者和专业人员进一步的讨论和反思。

沙永杰

2012年3月24日

any more, but right a threshold for architecture and urbanism in Asia in near future. For such, this book has raised a proper issue in a right time for Asian context.

Some small Asian cities and towns have touched me even greater than those Asian global cities, by their traditional ways of life, bonding with past, nature and happiness of people. I do believe that will not be fully erased by globalization or any other force. On such places, what often emerging in my mind is to question how such situation in China is, in an immense scale. Unfortunately I know it little and there is little information available too. We are too busy providing blueprints for such places to know well about them. Rapid and large-scale urbanization is inevitable in China this period. It is a challenge as all know, and a great opportunity as well referring to the developing of modernism and modernism city planning in West after WWI and WWII. Personally I think sufficient self-understanding is prerequisite for defining adaptive solutions for Chinese architecture and urbanism, and looking at West for relevant experience at some historic periods makes more sense than following the state-of-the-art thoughts and fashions nowadays. Both will possibly lead to a dialogue indeed between East and West.

This book is the first one in a newly launched book series titled REFLECTION by Tongji University Press. The series aims to seek seeds of thoughts behind phenomena and forms. Hyunjoon Yoo's book will play a positive role well for such aim, and key terms of modernism, East and West in this book will definitely draw further debates and rethinking by Chinese architects and urban designers.

Sha Yongjie
March 24, 2012

亲爱的中国读者：

大多数的世界历史是西方历史学家们写就的，建筑历史也不例外。世界建筑史就是西方建筑史学家的作品。我学生时代学习的建筑史也都是欧洲人的观点。16世纪之前，欧洲与亚洲的文化在彼此隔绝的状态下发展，因此以欧洲视角理解16世纪前的建筑历史无可厚非。然而，一旦两大洲之间开始广泛的文化交流，追寻文化嬗变的缘由及影响变得异常复杂。从这方面讲有必要对世界建筑进行多角度的审视。20世纪的建筑史也来自欧洲，现代主义建筑史被描述为基于工业革命，以及由此带来的材料与技术的更新。以工业革命解读现代主义建筑史令人信服，但正如前面所说，我们也需要从另一个视角理解现代主义建筑发展过程中的历史事件。我希望能够立足东方，从文化交流的视角而不是如西方人那样仅以工业革命和技术进步来诠释现代主义建筑史的成因。

从欧亚文化交流的角度来分析，中国文化在亚洲文化发展历程中扮演了极其重要的角色。这也就是我为什么非常乐意与中国读者分享我的观点的原因。我在哈佛大学读书时，以论文的形式完成了本书的初稿，有些教授支持本书的观点，有些教授则不以为然。我非常期待中国读者对这一观点的看法，也希望能够抛砖引玉，引发对现代主义建筑史的再认识。

最后，我想感谢江岱在过去三年里为这本书的出版所付出的努力，她将我的观点精心地翻译成中文。感谢沙永杰教授的推荐，感谢同济大学出版社！没有他们的帮助，这本书将无缘与中国的读者见面。希望中国读者能够喜欢这本书，并与我分享你们的观点。期待21世纪的世界文化更和谐！

俞炫准

首尔，2012年2月

Dear Chinese Readers,

Most of world history has been written by western historians. Architectural history is not an exception. World architectural history has been written by western architectural historians. As a student, I studied architectural history from European's view point. Europe and Asia have separately developed their culture until 16th century and there is no big conflict in reading architectural history before 16th century from European's viewpoint. Once the two continents started to exchange their culture extensively, however, it became more complicated to find out the cause and effect of cultural evolution. In this regard, world history has to be reviewed from various viewpoints. The history of the last century was written by the European viewpoint, and as a result, the modern architectural history was written based on the Industrial Revolution and the material and technological changes caused by it. It is quite a convincing idea to read modern architectural history from the viewpoint of Industrial Revolution. But, as I mentioned, we need another viewpoint to read the events of modern architectural history. I suggest the interpretation of the modern architectural history from the Eastern point of view and from the viewpoint of cultural exchange instead of the industrial and technological revolution created by Europeans.

In terms of cultural exchange between two continents, Chinese culture has played a key and the most important role in Asia throughout history. In this respect, I am very pleased to share my idea with Chinese readers. When I first wrote first draft of this book for a thesis at Harvard University, some professors did like its new concept and others didn't. I'm anxious to see Chinese readers' response. I hope this book will help the readers to develop their own viewpoints on modern architectural history.

Lastly, I would like to thank Jiang Dai for her enormous efforts to publish this book for the last three years. She elaborately translated this book to convey my ideas in Chinese. I also thank Professor Sha Yongjie for introducing me to Tongji University Press and Tongji University Press for publishing this book in China. This book was not able to be published in China without their help. I hope that the Chinese readers will enjoy this book and share their ideas with me. Let's make the culture of the 21st century better together.

Hyunjoon Yoo
Seoul, Feb. 2012

前言

1999年的冬天，我在纽约的一家建筑师事务所工作，萌生了写这本书的念头。当时在纽约读书和工作的韩国建筑师们每周五晚都会聚在一起，讨论彼此感兴趣的话题。有一次，一位同学做了日本传统建筑方面的演讲，而下一周讨论的主题是柯布西耶的带形窗。尽管两次讨论的主题完全不同，但我感觉它们所描绘的空间具有某些共性。特别是密斯的图根哈特住宅的空间序列与日本传统建筑几无差别。从那时起，我开始带着东方的视角，研究西方现代主义建筑大师们的建筑作品。

这本书看上去似乎靠不住，因为它的逻辑推理出于直觉，看似荒诞。而我既没有建筑史的博士学位，也没有受过文化人类学的专业训练，我只是学过设计的建筑师。这本书更像是我在学习和工作中的思考。这些思考源自我的宗教信仰、建筑理念、亚洲人的身份，以及通过在美国的学习获得的西方视野。

本书旨在追寻现代主义建筑的血统与宗谱。在西方建筑史中，之所以在当代出现一种全新的建筑空间，势必是艺术家们受到一种堪称"病毒"的思想的侵入。本书正是我在寻找"病毒"性的现代主义建筑源起过程中的一个副产品。

希望读者能够喜欢这本貌似荒诞的小书，能够理解我对现代主义建筑的"有趣"解读，毕竟它提供了一个解读建筑风格的新视角。

感谢我的论文导师——哈佛大学的玛丽安·汤普森教授，哈佛大学的安德烈娅·莱丝教授和我在MIT的论文导师长仓威彦教授和已故的威廉·J.米切尔教授。最后，我将这本书献给给予我全身心支持的父母、妻子和两个儿子，并感谢上帝对我的垂爱。

俞炫准

2008年2月

Preface

It was the winter of 1999 when I was working for an architectural firm in New York that I first conceived the main idea of this book. At that time, the Korean architects who had studied in the United States and were working for architectural firms in New York got together every Friday night to give presentations and discuss topics of mutual interest. One day, someone gave a presentation on Japanese traditional architecture, and the next week's topic was the ribbon windows of Le Corbusier. Even though the two topics were totally different, I felt some kind of relationship between the characteristics of the two spaces they depicted. In particular, the spatial sequence of *Tugendhat House* by Mies van der Rohe and Japanese traditional architecture looked exactly the same. Since then, I studied the works of the modern architectural masters in the West from an Eastern point of view.

This book is like a house made of cards. Since the logical steps of the book were connected intuitionally, its composition could look absurd. I do not have a Ph. D. in architectural history or cultural anthropology. I am just an architect who designs. This book is, therefore, more like a summary of my thoughts when I studied architecture at school and worked for an architectural office. These thoughts are the result of my religious faith, my thoughts about architecture which is my vision, an Eastern point of view formed due to my background as an Asian, and a Western perspective obtained when I studied mostly in the United States.

The aim of this book is to pursue the hereditary genealogy of modernism architecture. In the history of Western architecture, new architectural space appeared in modern times. There must be a kind of "virus" that affects the thoughts of artists when new things appear. This book is a byproduct created in the process of searching for the origin of the modernism "virus."

I hope that readers enjoy this book, and understand it, even when it might seem a little "absurd." I would appreciate it if the readers think, "This architect interpreted architectural history in an interesting way." Anyway, a creation of new things in architecture or design starts from a fresh point of view.

I would like to thank Prof. Maryann Thompson, my thesis advisor at Harvard University, Prof. Andrea Leers at Harvard University, the late Prof. William J. Mitchell and Prof. Takehiko Nagakura, my thesis advisor at M.I.T. Finally, I would like to dedicate this book to my parents, my wife and two sons, all of whom have devotedly supported me; and the God whose hand is invisible guiding my life.

Hyunjoon Yoo
Feb. 2008

目录

Table of Contents

引

Introduction

现代主义在20世纪的影响之大，使得人们在谈及那一段时期的文化时无法回避它。至今，人们认为现代主义是工业革命和随之而生的功能主义所引发的一种思潮。在本书中，笔者则试图偏离固有的认知角度，把现代主义理解为一种文化的变异：现代主义孕育于东方文化传入西方的过程中。21世纪的人们已经自然而然地接受了“地球村”的概念，此时再来划分东方和西方或许显得落伍，但回顾东西方之间过去已经发生的文化交流及其所留下的印迹，对于生活在21世纪、身肩建设后现代时代之任的我们来说，是极有意义的尝试，而本书的写作正是秉持了这一观点。以下为便于记述，把包括中国、韩国、日本等在内的东亚地区统称为“东方”，把欧洲国家称为“西方”。

Modernism wielded such a great deal of influence in the 20th century that the culture of that century cannot be discussed without referring to it. So far, modernism has been considered as a trend of thought that emerged from the Industrial Revolution and the functionalism that followed it. This book, however, intends to deviate a little from the existing point of view and interpret modernism as a sort of a cultural variant that came into being when the Eastern culture was introduced to the West. It sounds somehow obsolete to draw a distinction between the East and West in the 21st century, when the term "global village" is naturally accepted. However, this book was written from the perspective that to trace the preexisting cultural exchanges between the East and the West could be a meaningful exercise for the people of the 21st century who are leading the post-modernism era. For the purposes of this book, Far East Asia – including China, Korea and Japan – will be referred to as "the East" and the European countries will be referred to as "the West."

图1
万神庙（118—128年，罗马）

Fig. 1
Pantheon, A.D.118-128, Rome

文化的产物——建筑

在互联网和传媒发达的21世纪，大部分人即便没有亲自去过国外，也能接触到来自其他国家的信息。随着信息交流的愈益频繁，为亲眼目睹和体验图片或文字中的事物而进行旅游的人变得越来越多，这是信息技术相对落后的时代无法比拟的。那么，这些游客到国外最常做的事是什么呢？恐怕就是站在当地著名建筑物前拍照留念了。一个国家的文化往往体现在建筑上，没有一种人类行为像建筑这样需要投入巨额的资金并耗费长久的时间。建筑是人类智慧和意志的结晶，它代言一个时代的文化，并且跨过岁月的长河，流传后世。

那么，建筑是如何让不同时代的人们跨越岁月的长河进行沟通的呢？实现沟通的媒介便是建筑空间。与绘画或音乐不同，建筑有其独特的沟通媒介——“虚

Architecture as a Cultural Product

In the 21st century, with the exponential growth of the Internet and media, people can access images and information about other countries without actually going there themselves. With the increasing exchanges of information, however, the number of travelers who want to directly see and experience what they have seen or read about on the Internet or other media is increasing greatly, compared to the time when information technology was undeveloped.

One important thing that many travelers do when they visit foreign countries is to take pictures in front of famous architectural buildings of those countries. This is because the essence of a country and its national culture is embodied in its architectural buildings. No other human behavior consumes as much time and money as architecture does. The combination of the wisdom and will of human beings create the outcomes of architecture. Architecture represents the culture of

空”(void)。绘画或雕塑作品可以实现由“实体”(solid)的物所传达的象征性，建筑上的和谐与协调也正是音乐所具有的特点，唯独虚空的空间所提供的三维信息不存在于上述任何一种媒介中。因此，虚空的空间是建筑文化最主要的沟通媒介，并构成了多种不同的建筑文化特色。因此，虚空空间的设计，是界定文化特性的一种重要手段。

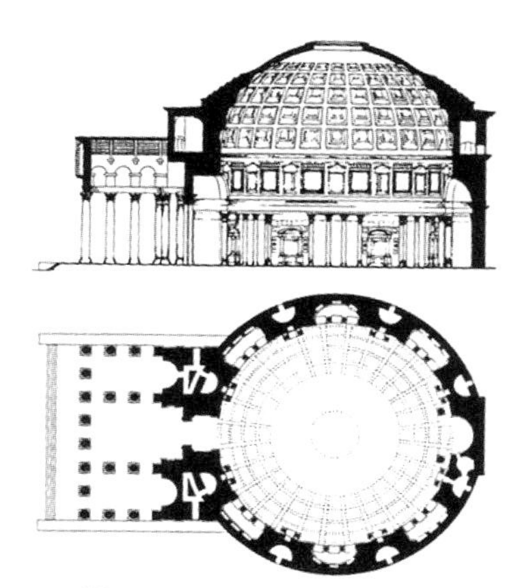

图2
罗马万神庙的平面与剖面

Fig. 2
Pantheon. Plan and section

a certain era, which is inherited by succeeding generations.

How, then, can architecture facilitate communication among people of different times? The medium that makes that communication possible is architectural space. The medium of communication that only architecture possesses, unlike paintings or music, is the "void." The symbolism that solid objects represent can also be found in paintings or sculptures, and the rhythm and harmony in architecture is also a characteristic of music. The 3-Dimensional information that is given from vacant space, however, does not exist in any other medium. In this regard, void space is the main medium of communication and feature of architectural culture. Consequently, the way in which void space was designed could be an important means of defining the characteristics of a culture.

光的存在条件——虚空

太初，上帝创造了天地。大地一片混沌，空无一物，黑暗笼罩着渊面，上帝的灵在水面上运行。上帝说："要有光"，就有了光。

《创世纪》[1]第一章第一句至第三句

《圣经》的这一段话清楚地告诉我们上帝创造虚空与光的次序。其中，"空"在英文的《圣经》中表述为"void"，说明上帝创造天地时，先创造空间，然后创造了光。光是认知世界的最优先的条件，但若没有虚空，光也无法存在。那么，虚空又是如何被认知的呢？

Void, a Prerequisite of Light

"In the beginning, God created the heaven and the earth. And the earth was without form, and void; and darkness was upon the face of the deep. And the Spirit of God moved upon the face of the waters. And God said, Let there be light: and there was light."

Genesis 1:1-3 (KJV[1])

In this verse, the order of the creation of the void and light is well explained. It is clear that God made the void first and then light when he created the world. Light is an entity that is needed first when recognizing the world, but that light cannot exist without the void. Then, how can we recognize the void?

---

[1] KJV为英王詹姆士一世钦定版圣经（King James Version of the Bible），1611年。这是圣经最传播最广、也最为根本的一个版本。

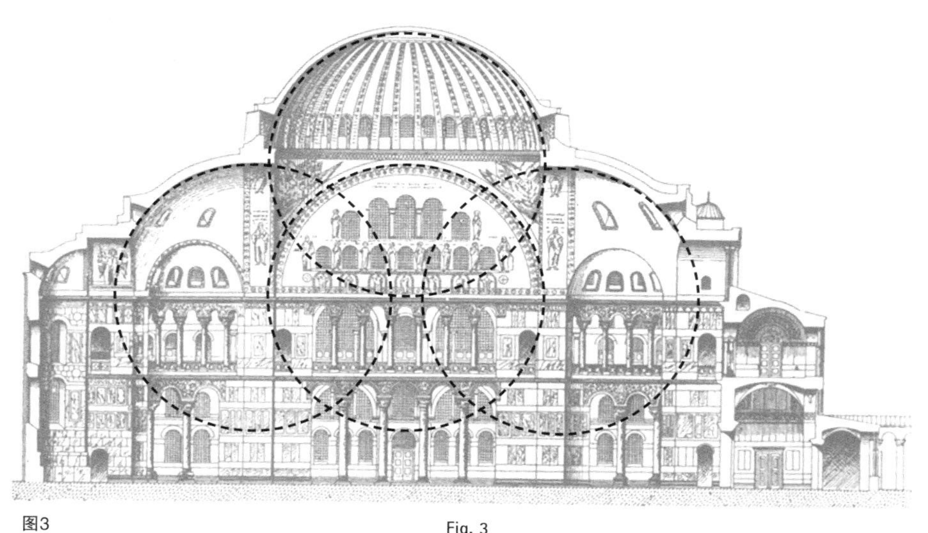

图3
圣索菲亚大教堂（君士坦丁堡，532—537年）

Fig. 3
Hagia Sophia, 532-537, Constantinople

人的空间知觉

人是一种三维存在。换句话说，人具有一定的体量。三维的人能充分地认知二维平面或一维的点。要充分认知三维空间，只能通过四维或更多维的参照。三维的人要认知三维的空间，需要第四维——时间的帮助。眼球搜集到二维信息并将其传递到视网膜。随着时间的推移，搜集到越来越多的二维信息后，人利用意识这个“软件”，将信息重新组织成三维的空间。当然，在这个过程中，由于双眼处在两个不同的位置上，它们能获得两种不同的信息，因此即使没有时间这一维的帮助，人也能获得很强的空间感。此外，在声音和光影等基本信息的帮助下，结合过去的经验，空间感是可以被逐渐培养出来的。

认知虚空，必须要有实体。丹麦格式塔心理学家爱德加·鲁宾（1886–1951）

Human Perception of Space

Human beings are a form of three-dimensional existence. That is, they are a form of existence that has a volume. Thus, three-dimensional human beings are able to fully recognize one-dimensional points or two-dimensional planes. Three-dimensional space is completely recognizable only by an existence which is four-dimensional or higher. If human beings want to recognize three-dimensional void space, they need the help of four-dimensional time. Man gets two-dimensional information that is delivered to the retina, collects that various information gathered in different times, and then reorganizes it to create three-dimensional space by using "software", which is one's consciousness. In this process, man also can get a great deal of spatial sense without the help of time by receiving two kinds of information through two eyes in different positions. Besides this, people can develop spatial sense all the more by using their past experiences with the basic information, such as sounds or shadows.

在1915年画的视觉认知图《面孔和花瓶》（图4）很好地说明了“虚”与“实”相互依存的关系。在这幅画中，为了看到白色的花瓶，人们需要两个黑色的面孔充当背景；反过来也是如此。正如感觉光线需要阴影，认知虚空需要有实体；反之也可以说，一旦创造出了实体，虚空也就自然形成了。

表面上看，人类的建筑行为是在建造实体，但其实是在创造人类可以使用的空间——虚空。虚空的空间能反映出设计者的思考和哲学。

图4
《面孔和花瓶》

Fig. 4
Face and Vase

The solid is absolutely needed to recognize the void. The principle is seen in Figure 4, a cognitive optical illusion created by Danish Gestalt psychologist Edgar Rubin in 1915. This picture, *Face and Vase*, shows an interdependent relationship between the solid and the void. In the picture, the black background that depicts two profiles is necessary to recognize the white vase, and vice versa. In the same way that a shadow is needed to see light, a solid is needed to notice the void. If we reason the other way around, the void is created as a byproduct when there is a solid.

The superficial purpose of human beings' architectural behavior is to create something solid, but the ultimate purpose is to create the void that man can use. Void space reflects the architect's thoughts and philosophy.

东西方的交流

东西方交流的最早记录可追溯到中国的汉朝和欧洲的罗马帝国时期。在当时，中国商人带到欧洲的丝绸，成为罗马人狂热追求的对象，不惜用等量的黄金换取。但东西方思想上的交流则直到16世纪左右才开始，天主教传教士们翻译了孔子和老子的著作，并将其引入西方。

在现实生活中，西方进口中国的瓷器，间接促进了双方的文化交流。使用笨重的金属餐具的西方人，将瓷器视为艺术品。瓷器上描绘的图案，使得西方人自然而然地了解到中国人的生活情景。中国文化的影响在欧洲贵族阶层迅速扩散开来。在双方展开大规模的交流之前，东方和西方两个世界，由于距离遥远等原因，在完全不同的哲学基础上，各自发展自己的文化长达2000多年。

Exchanges between the East and the West

The first mentions in literature of exchanges between the East and the West appeared during China's Han Dynasty and the Roman Empire in Europe. At that time, Chinese merchants went to Europe with silk, and the Romans who were fascinated by the silk exchanged the silk for gold of the same weight. The spiritual exchanges started from about A.D. 1500 when the books of Confucius and Lao Tzu, translated by Catholic missionaries, were introduced to the West.

In real life, as Chinese tableware, a state-of-the-art item from the point of view of Westerners, who used heavy metal tableware at that time, was exported to the West, the Chinese life depicted on the chinaware was naturally introduced to the West, and familiarity with Chinese culture became widespread among European noblemen. Until this kind of exchange happened on a large scale, however, the two worlds had built their own cultures based on totally different philosophies for

本书将比较和分析在东西方两种文化环境中虚空空间的性质，试图描绘出东西文化交流在建筑领域所产生的影响。两种文化中的虚空空间究竟有什么不同？西方建筑有遵循几何学规则的倾向，东方建筑则重视组成元素的相对关系，这是它们之间最大的不同点（图2，图3，图5，图6）。

东西方文化有着如此不同的建筑空间，自16世纪开始全方位的交流后，逐渐

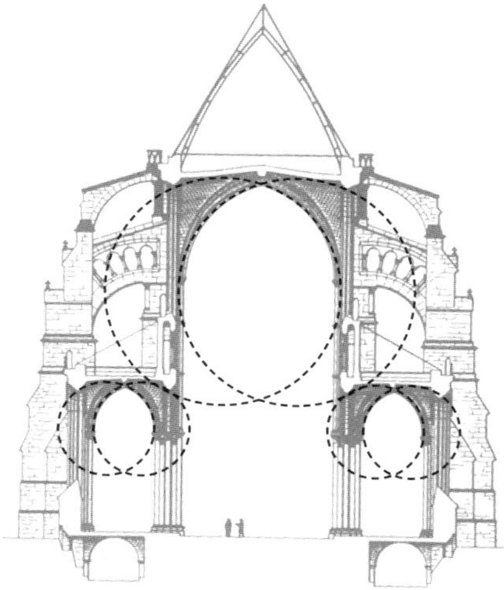

图5
圣母大教堂（夏特尔市，哥特式，1194—1230年）

Fig. 5
Cathedral Notre-Dame, Chartres, 1194-1230.
High Gothic Style

nearly 2,000 years, due to geographic and other reasons.

This book aims to determine the influence that the cultural exchanges between the East and the West exerted on architecture, by comparing and analyzing the characteristics of void space in the two cultures. What is the difference between the voids in the two cultures? The biggest difference is that whereas the architectural buildings in the West are likely to be designed in accordance with geometric rules, the void in the East is made according to the relationships among architectural elements (Fig.2, Fig.3, Fig.5, Fig.6).

The two cultures, which had displayed different styles of void space, gradually

发生了变化。西方开始出现具有东方特色的建筑和空间；东方则受到西方科学技术的影响，出现了实用主义思想。中国对西方的影响最初体现在英国的风景式园林，此后又对密斯·凡·德·罗和勒·柯布西耶设计的建筑空间产生影响。这两位著名的西方现代主义建筑大师善于运用东方建筑空间最重要的特质——“相对性虚空空间”。此后的现代主义建筑代表人物路易·康（1901–1974），把西方传统建筑的几何图案组合与东方建筑所具有的禅意美学结合起来，创造了独特的现代主义建筑风格。

在东方，尽管也深受密斯和柯布西耶的影响，却没能继承自己原先拥有的东方价值，只接受了钢筋、水泥、平屋顶等表面形式和新技术，错失了近代东方建筑进一步发展的好机会。此后，日本建筑师安藤忠雄（1941– ）在日本经济发展、

changed from the 16th century onward, when full-scale cultural exchanges began. While the Eastern style architecture and space started to be manifested in the West, the utilitarianism influenced by Western science and technology appeared in the East. The Chinese influence on the West is reflected in the picturesque style of garden design in England, and later on in the architectural space designed by Mies van der Rohe and Le Corbusier. These two famous Western modern architects adopted "relative void space," the most important feature of Eastern space. Louis I. Kahn, the next successor of modern architecture created his own modern architectural style by combining the geometrical space of Western traditional architecture and Zen-style space of the East.

By contrast, the East adopted only the superficial and technical contents of Western architecture, such as iron, concrete, or flat roof, while it is greatly influenced by Mies and Corbusier. It failed to succeed to its inherent Oriental values to its

政局稳定的环境下，成功地将西方建筑的几何形体与东方建筑空间的禅意空灵融合在一起，在20世纪后期的现代建筑史上，写下浓重的一笔。

现代主义是东方文化的比较哲学传入西方后形成的文化杂合。换句话说，东方文化的传入使西方人的思维方式发生了变化，使几何图形的、固定的西方建筑空间，发展成相对的、流动的灵活空间。以下将回顾两种文化创造的建筑空间，探寻能够证明上述假设的线索。

图6
布雷设计的牛顿纪念碑（1780年代）。
这是一个直径170m的球形空间。
几何空间的设计传统历时弥久。

Fig. 6
Étienne-Louis Boullée, *Cenotaph for Sir Isaac Newton*, 1780s. This space has a sphere-shaped void 170m in diameter. Geometric void space has continuously been created throughout time.

regret, and the Eastern modern architecture missed a good opportunity to make a leap forward. Later, Japanese architect Tadao Ando achieved something remarkable that marked an epoch in the development of modern architecture in the late 20th century by combining the geometric architectural style of the West and Taoist space of the East on the basis of the economic development and political stability of Japan.

Modernism is a cultural hybrid that appeared incidentally when the value of relative philosophy of the Eastern culture arrived in the West. As the cultural inflow from the East brought about an ideological paradigm shift in the West, geometric and fixed void space of the West developed relative, fluid, and flexible space. This book seeks to trace the evidence that can support the abovementioned hypothesis by studying the genealogy of void space in the two cultures.

两个不同的世界

Two Different Worlds

古代伟大的思想家们

东西方文化具有截然不同的精神世界；同样，两种不同文化中的建筑空间也有着很多不同点。为了解这种差异的背景，我们首先要了解两种文化的思想根源。有趣的是，那些奠定东西方文化根基的伟大思想家们所处的时期相近，都在公元前570年至公元前300年之间。

这一时期，数学之父毕达哥拉斯（约前570–前495）、建立西方哲学基础的柏拉图（约前427–前347）、几何学之父欧几里得（约前300），在西方建立了自己的学术体系。而在东方，老子（约前571–约前471）、孔子（前551–前479）、释迦牟尼（前563–前483）奠定了东方思想的基础。

Great Ancient Philosophers

Since the different worlds of mind created different void space in the East and the West, a study on their ideological origins is needed to understand void space of each culture. The interesting fact is that great philosophers who laid the foundation of two worlds of mind lived during a similar period from around 570 to 300 B.C.

Pythagoras (ca. 570 – ca. 495 B.C.), the father of mathematics, Plato (ca. 427 – 347 B.C.), who formed a foundation of Western philosophy, and Euclid (fl. 300 B.C.), the father of geometry, in the West; and Lao Tzu (571? – ? 471 B.C.), Confucius (551 – 479 B.C.), and Buddha (563 – 483 B.C.) in the East laid the cornerstones of the realm of thought.

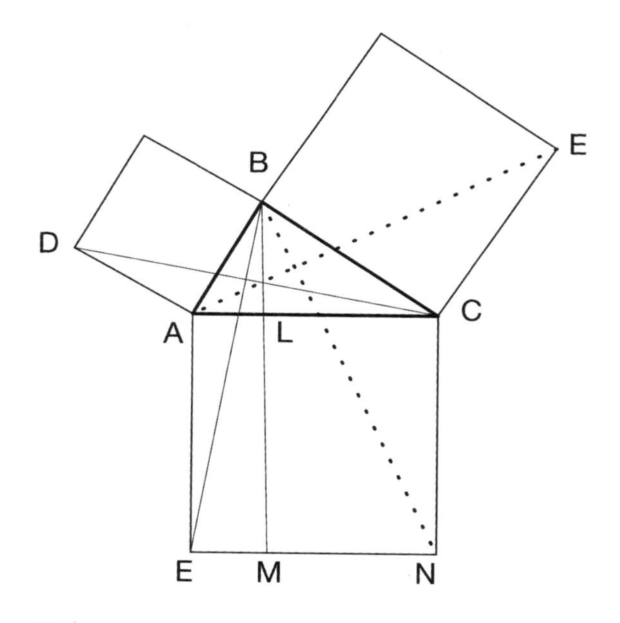

图7
毕达哥拉斯定理（勾股定理）

Fig. 7
Proposition of Pythagoras

柏拉图在其老师苏格拉底（前469-前399）去世后，环游诸国，结交了许多著名的数学家。在古希腊殖民地昔兰尼，他向特奥多罗斯学习几何学；在埃及，他与毕达哥拉斯学派的成员形影不离。这种经历使他与蔑视数学的苏格拉底不同，他热爱数学。当他在阿卡德摩斯城郊创立学园[2]（今称学院、大学、研究院等），拒收没有数学（几何学）基础的人。在“洞喻”中可以看到堪称柏拉图哲学核心思想的“理念”论。他认为，世界上的所有现象，都是“理念”的影子，人通过五感（视觉、听觉、嗅觉、味觉、触觉）认知的只是本质——“理念”的表象，而非实际。这种理念思想和背后的数学式思维，成为欧洲精神世界的基础。

在希腊思想传播至整个欧洲的过程中，宗教起了重要作用。早期基督教传入罗马和地中海地区时，多少具有希伯来文化的色彩，但随后在希腊文明的框架内

Plato formed close friendships with famous mathematicians when he traveled through many countries after his teacher Socrates died. He learned geometry from Theodoros in Cyrene, and kept company with Pythagoreans in Egypt. Influenced by them, he began to like mathematics – unlike Socrates who despised it. He would not even admit students without a basic knowledge of mathematics (geometry) when he founded the "Academy" school in the grove of Academos.[2] His "allegory of the cave" reveals the theory of "Ideas", idea of Plato's philosophy, which considers all the phenomena in the world as mere shadows of Ideas, and the things we recognize through our five senses as not real things, but phenomena of Ideas, which is the essence of phenomena. The theory of Ideas and the mathematical thought to search for that became the basis of the European mind.

Religion played an important part in the dissemination of Greek thought through-

---

[2] **Kim, Yongun.** ***Interesting Stories of Mathematics*****. Seoul: Seohaemunjip, 2005.**

得到重组。基督教的重组始于出生于希伯来文化和希腊文化交汇地——基利家省塔尔苏斯的圣徒保罗（5-67），此后经圣奥古斯丁（354-430）而大功告成。

圣奥古斯丁在整理早期基督教教义时，借助柏拉图理论的形式以帮助人们理解。因此，基督教和柏拉图哲学有很多相似之处。

比如，《圣经》中《约翰福音》第一章第一篇写道："太初有道，道与上帝同在，道就是上帝。"这里所说的"道"是指"逻各斯"。《约翰福音》第十四章第六篇还说："我是道、真理、生命；若不借着我，没有人能到父那里去。"可见，逻各斯就是耶稣的化身，若没有耶稣或逻各斯，无法到达与理念类似的拥有绝对价值的天国。

out the entire Europe. When early Christianity with Hebrew features spread to Rome and the Mediterranean region, it was reorganized on the basis of Greek culture. This task originated by Apostle Paul (A.D. 5-67), who was born in Tarsus of Cilicia, the borderland of Hebrew and Greek culture, and accomplished by St. Augustine (A.D. 354-430).

St. Augustine used the form of Plato's theory to help people understand the doctrine of Christianity when he arranged the religious theories of early Christianity, and that is why there are many things in common between Christianity and the philosophy of Plato.

For example, in the Bible verse "*In the beginning was the Word, and the Word was with God, and the Word was God* (*John* 1:1)," the "Word" means "Logos." The verse, "*Jesus saith unto him, I am the way, the truth, and the life: no man cometh unto the*

“逻各斯”在字典里的定义是“了解事物存在的、人的理性”，理性是通往天国或理念等理想世界的道路。这种思想似乎也是受到宗教的影响。

综上所述，西方人普遍的思维方式是，存在绝对真理的世界，而合乎逻辑的逻各斯是通往那一世界的唯一方法。正是因为这种思维方式，数学才有可能成为一门对西方文化具有重大影响的学问，并在此基础上实现了科学革命。在将数学上升为一门学问的人中，最具代表性的人物就是希腊数学家、宗教家毕达哥格拉斯。最初使用“philosophic”[2]和“kosmos”[3]两个单词的人就是毕达哥拉斯。毕达哥拉斯和其学派最早开始研究弦乐器的琴弦长度和音高的关系，他们认为数学的和谐是通往绝对真理的道路。

Father, but by me (*John* 14:6). indicates that "Logos" was incarnated into "Jesus" and without Jesus, or Logos, people cannot reach heaven, which has an absolute value, as Ideas do.

The definition of "Logos" in the dictionary is "man's reason to understand the existence of things". It means the road to heaven or an ideal type, such as an Idea, can be reached by reason. In this regard, the Western thought that we can reach Ideas only by reason seems to be influenced by religion, too.

As seen so far, the overall underlying way of thinking in the West considers that there is a world of absolute truth, and the road to get there is reached only by logical Logos. As a result, mathematics had a great influence on Western culture,

---

[2] 原意为“自然科学”，后被引申为“哲学的”

[3] 原意为“思想完整的一统体系”，后被引申为“宇宙”

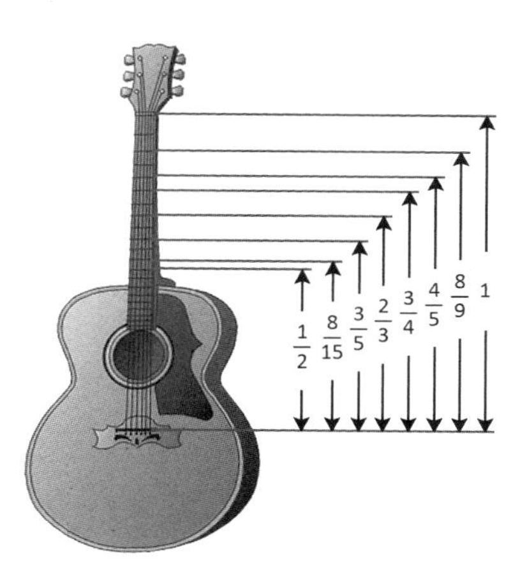

图8
毕达哥拉斯以数学定义的和谐

Fig. 8
The mathematical definition of harmony by Pythagoras

and the science revolution was made possible. Pythagoras, a Greek mathematician and religionist, is one of the people who raised mathematics to the rank of science. It is he who first used the words "philosophic", which means philosophy, and "kosmos", which means universe. Pythagoras and his school also studied the relationship between the length of the string of stringed instruments and sound for the first time, and they seemed to think that mathematical harmony is the road to absolute truth.

Plato was influenced by Pythagoras through friendship with the Pythagoreans

柏拉图在结交毕达哥拉斯学派，确立自己哲学思想的过程中，深受毕达哥拉斯的影响。此后，欧几里得集毕达哥拉斯和柏拉图之大成，写出了《几何原本》。经过这一系列的传承，“和谐”、“数字”、“理性”和“几何”成为西方文化的关键词，数学进而对整个西方文化发挥了极其重要的作用。西方文化的内核，人们相信凡神圣或理念都是与现实生活相分离而存在，通过数学可以接近真理。而这种世界观完整地贯彻到了建筑空间。纵观西方近代以前的所有宗教建筑，能够很容易地看出，其空间结构是根据数学的分析来设计的。

当西方人基于数学的逻辑，形成客观而绝对的价值观的时候，东方则形成了相对的价值观，即重视“关系”的文化。简单地说，东方人不相信绝对的善。东方也有与西方的天国相类似的世外桃源这样的场所概念，但仔细分析孔子的哲学可以发

when he established his own philosophy. Afterwards, Euclid wrote *Stoicheia (Elements)* based on the thoughts of Pythagoras and Plato. In this process, words such as "harmony", "number", "reason" and "geometry" became key words of Western culture, and mathematics exerted a great deal of influence on Western culture in its entirety. At the heart of Western culture, there is a belief that something divine, or Ideas, exist apart from the real world, and people can get close to the truth through mathematics. This world view is revealed as it is in the architectural void space. If we take a look at all the religious architecture in the West built before the modern period, we can easily find out the fact that the spatial compositions were designed according to mathematic analysis.

While an objective and absolute point of view was shaped on the mathematic logics in the West, the culture which considers a relative point of view or a "relationship" as important was formed in the East. There is no belief in the absolute

现，东方人更重视相对的价值和关系。孔子视“中庸”为最高道德准则，它不属于绝对的善，而是可以随着情境和关系发生变化的德。相反，西方文化基于个人主义，依靠演绎的思维发展了科学。从笛卡儿所说“我思故我在”中可以看出，西方哲学根源于独立的自我。因此，在物理学方面，从古希腊时代到20世纪初期的现代物理学，西方人都在探寻“分子”、“原子”、“夸克”这种构成宇宙的最小单位。

东方的“德”与西方不同。比如，东方人认为“孝”是道德最基本的因素，它基于父与子二者之间的相对关系。孔子教导人们在相对的关系中达到“善”，而不是在绝对的道德规律中达到最高境界。中国古代的哲学——阴阳论，是东方相对主义思想最典型的例子。“阴”和“阳”不是两个对立的“能量”互相冲突的关系，而是相互依存，努力成为一体的关系。

virtue in the East. Even though there is a spatial concept of the "Happy Valley", which is like the heaven in the West, the Confucian philosophy, which can be said to have built the basis of Eastern thought, puts importance on relative values and relationships. Confucius considered "moderation" as the most important virtue, and it is not an absolute concept of virtue, but a virtue that changes according to the situation and relationship. On the other hand, Western culture is based on individualism, and science has developed on the basis of deductive thinking. As Descartes said "I think, therefore I am", the Western philosophy is based on an independent self. That's why physics searched for the minimum unit that constitutes the universe, such as "molecule", "atom" and "quark" from the Greek period to the early 20th century.

Let us take a look at the Eastern virtue that developed differently from that of the West. For example, "filial piety," which is considered as the most fundamental

四灵兽图中左半部分（代表北方）的玄武（图9），乌龟和蛇看似在相斗，也有说是在交尾。可见，东方的价值观不是单纯、绝对的二元思维，而是建立在相对关系的基础之上。

除了相对价值观外，东方文化的又一重要因素是“无”。在印度，数字“0”并不意味着什么也不存在，而是意味着存在人还无法认知的数字。同样，东方文化中的“空”，不是简单的否定，仅仅说明没有物质的存在，而是有进一步的象征意义。强调“空”的价值的东方哲学家当数老子。他说，触手可及的物质过多，潜在的成长的可能性反而会消失。这种思想很好地体现在老子《道德经》的第十一章中：

三十辐共一毂，当其无，有车之用。埏埴以为器，当其无，有器之用。
凿户牖以为室，当其无，有室之用。故有之以为利，无之以为用。

moral virtue in the East, is based on the relative relationship between parents and their children. Instead of teaching an absolute moral code, Confucius tells us to reach the good through relative relationships. The most certain example of relative thought of the East can be found in the principle of Yin and Yang in ancient Chinese philosophy. Yin and Yang, the opposite "energies" do not have an antagonistic relationship in which they conflict with each other, but are considered as an interdependent relationship in which they strive to become one.

In the *Picture of the Black Tortoise* (Fig.9) in the left part of the *Picture of Four Deities*, a tortoise and a snake look as if they are fighting, but actually they are copulating. The Eastern value system is not based on the simple, absolute, and dichotomous thought, but on relative relationships.

图9
玄武

Fig. 9
Picture of the Black Tortoise

Other than relative point of view, another important keyword of Eastern culture is "emptiness." In India, the number "0" is not "nothing," but "an existing, but unrecognizable number." The "emptiness" in the East has a further symbolic meaning, rather than simply negatively meaning that the thing does not exist. Lao Tzu, a philosopher who emphasized "emptiness" said that once the existence that we can grab in our hands is full, the potential to grow is reduced. His thought is well explained in his *Tao Te Jing* Chapter 11.

"Molding clay into a vessel, we find the utility in its hollowness; Cutting doors and windows for a house, we find the utility in its empty space. Therefore the being of things is profitable, the non-being of things is serviceable."

Lao Tzu. *Tao Te Jing*. Ch.11.

一般被解释为否定含义的“空”，以老子的观点看，却被视作百分之百的可能性。老子的这种思想体现在东方建筑空间，尤其是日本的枯山水和神社中。

枯山水不是常见的那种栽满树木的花园，而是留有空间的花园。神社则先划定两块空间，其中一块基地上修建神社，另一块基地则空出来留着，过了二十年之后，在空的一侧建新的神社，而把另一侧的拆掉。这样，空和实的循环以二十年为周期反复。东方思想的又一重要人物释迦牟尼也提倡空的意义，告诉人们心空即可入涅槃的境界。

综上所述，东方文化受孔子、老子和释迦牟尼的影响，价值体系的两个重要特征可以归结为“关系”和“空”。东西方价值体系的差异，从“空间”这个词语上便可略知一二。

From Lao Tzu's perspective, the void, which is usually interpreted as having a negative meaning, is seen as a state of 100% possibility. Lao Tzu's thought is reflected as it is in an Eastern architectural space, such as the Zen Garden and Shinto Shrine in Japan.

Whereas ordinary gardens are filled with trees, the Zen Garden is designed to be an empty space. In the case of the Shinto Shrine, there are two identical grounds. When a new building is built on one site, the building on the other site is demolished to be empty space. After 20 years, the new shrine is built on the empty site, and the existing shrine is demolished. By this process, the circulation of filling and emptying is repeated every 20 years. Buddha, another important figure in the world of Eastern thought, also taught that we can enter into the state of Nirvana by emptying our minds.

space
空間

图10
东西方“空间”的文字

Fig. 10
Words that means "space" in the East and West

As seen so far, the value system of Eastern culture can be summarized in two key words of "relationship" and "void," which were made under the influence of Confucius, Lao Tzu, and Buddha.

在西方，空间为“space”，“space”还指宇宙。宇宙的英语单词有universe、cosmos、space，三个单词通用。再仔细观察，cosmos是chaos（混沌）的反义词，意指具有数学的规则。从中可以看出，在西方人的头脑中宇宙、空间、包含数学规则的cosmos，相互有联系性，进而把空间视作“具有数学规则的空间”。

西方的“空间”是经过数学分析来建成的，而东方的“空间”是由空无一物的“空”和中间的“间”组成的单词，是“空”和“关系”的组合。因此，只看“空间”这个词就能了解到，东方人把“空间”看做是超过单纯的“空地”的、存在某种可能性的“空”，以及相对价值的“关系”。

The different value systems can be easily discerned when we compare the words that mean "space." In the West, the word "space" also means the universe. "Universe," "cosmos" and "space" are the words that express the universe. As the antonym of "chaos," the word "cosmos" implies that it has mathematical rules. With these examples, it is clear that the meaning of universe, space, and cosmos with inherent mathematical rules is interrelated in the minds of Westerners and they consider space as "a vacant thing with mathematical rules."

While the word "space" in the West is analyzed rather mathematically, the word "space" in the East is a combination of two characters that mean "emptiness" and "between." That is, the "space" in the East is a compound word of "void" and "relationship." It seems that Easterners understand space as "void," which has more potential than "simple emptiness" and "relationship," which has a relative value.

字母和汉字

前面讲到西方文化具有“数学秩序”的特点，东方文化的特点则是“关系”和“空”。东西方所使用的文字也体现出这一不同点。

英文字母的起源要追溯到埃及象形文字，其演变顺序是：埃及文字→西奈文字→腓尼基字母→希腊字母→拉丁字母。英文中共有26个字母，每个字母是固定不变的。现代物理学证明，物质和能量之间的界线是模糊的。传统的西方科学主张，物质是由最小单位“原子”组成的。原子是不能再分的、独立的最小单位。众多原子组成分子，分子的结合形成了人们肉眼看到的世界。字母的组成与传统的原子概念相似，不能再细分的26个字母按一定顺序排列组成单词，单词又组成语句。为字母赋予新的意义的方式是，将字母按一个轴线排列开来，

The Roman Alphabet and Chinese Characters

In the previous chapter, it was explained that Western culture has "mathematical order," and Eastern culture focuses on "relationships" and "meaningful emptiness." These features are found in the characters used in each culture.

The origin of alphabets goes as far back as the Egyptian hieroglyphics, and it seems to have been handed down with some changes in the order of Egyptian characters, Sinai characters, Phoenician characters, the Greek alphabet, and Latin alphabet. The Roman alphabet is composed of 26 letters, and neither of the letters changes. Whereas contemporary physics verified that the boundary between material and energy is ambiguous, material in traditional Western science was considered to be composed of the atom, the minimum decomposable unit. The atom was thought to be an unbreakable and independent unit. People thought that atoms compose molecules that constitute the world that we recognize.

只需要变换顺序就能组成不同单词。比如，D、E、N这三个字母，排列成“END”就是“结束”的意思，排成“DEN”就是“动物休憩的地方”（图11）。

中国的汉字，一个字可以加上不同的偏旁部首，形成不同意思的字，而且同一个偏旁部首，根据不同位置或长短，也可以组成完全不同的汉字。比如，一个的“一”和树木的“木”，用这两个字可以组成根本的“本”，表示否定的“未”，以及表示末梢的“末”三个字。也就是说，汉字可以根据偏旁部首笔画的不同组合组成不同的文字，表达不同的含义（图12）。

此外，西文中字母都是按一个方向线性排列的，但汉字可以在上下左右添加部首，具有多向性。如前面提到的例子，“一”可以加在“木”的上方，也可加到下方。也有很多汉字以左右、上下等结构组合起来，形成新的汉字。因此，汉字具有

The composition of the Roman alphabet is similar to the traditional concept of the atom. The unbreakable 26 letters stand in a row in the set order to form a word, and the words are combined to create a sentence. The way to make a new meaning using the alphabet is to arrange letters on an axis and change the order. For example, if there are three letters of D, E, and N, and the letters are arranged in the order of E, N, and D, it makes a word "END," and if the letters are arranged in different order of D, E, N, it makes a word "DEN," a place where animals take a rest.

In the case of Chinese characters, however, one character consists of some basic strokes and it can be made to represent a character with a different meaning. Even with the same basic strokes, totally different characters are made according to the position and the length of strokes. For example, the different position and the length of the strokes of "**一**" and "**木**" make three different characters; "**本**"

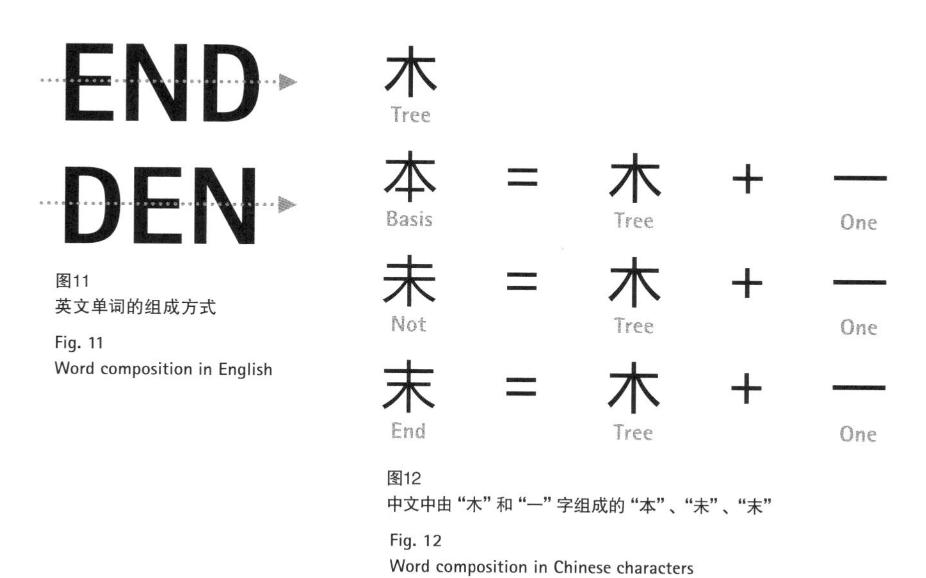

图11
英文单词的组成方式

Fig. 11
Word composition in English

图12
中文中由"木"和"一"字组成的"本"、"未"、"末"

Fig. 12
Word composition in Chinese characters

which means "basis," "未" which means "not," and "末" which means "end." That is, the meaning of a character is changed according to the interrelated relationship of position and length of some basic strokes that form a character.

Whereas Western alphabet letters are arranged in one direction, a Chinese character is made by adding strokes to every direction – the top, bottom, right, or left. As seen in the above example, "一" is added to the top of "木" or the bottom. Also in other Chinese characters, strokes are added to the top, bottom, right, or left to be mixed to create a letter with a new meaning. The freely growing pattern of the Chinese characters is reflected in Eastern architectural plans.

自由扩张的特性，而这种特点也体现在东方的建筑平面图上。

西方的宗教建筑或王宫，大多左右对称，按一个方向的轴线排列开来。但东方建筑的平面设计，根据周边环境灵活变化，并非左右对称，其结构更多是以自然增生的方式扩展。单一方向性和多方向性就是两种建筑文化最显见的不同。西方文化认为世界是神按照一定的规律组成的，东方文化则认为世界是关系的集合。两种文化的根本差异决定了两种建筑文化的不同，而这种不同点在国际象棋和围棋中也有表现。

While Western religious architecture or palaces are arranged on a one-directional axis based on bilateral symmetry, the plan in the East shows bilateral asymmetry as if the buildings are proliferating spontaneously conforming to the surrounding environment. Single direction and multi-direction are respectively one of the biggest architectural features of the two cultures, and they are due to the fundamentally different perspectives of each culture – Western culture considers the world to be a combination of rules provided by God, and Eastern culture sees the world as a sum of relationships. This difference is clearly seen in the favorite games of each culture – chess and Go.

国际象棋和围棋

游戏是反映文化特征的又一种媒介。国际象棋来源于名叫“恰图兰卡”的游戏，始于公元600年前后的印度。“恰图兰卡”于公元625年传入波斯，公元711年摩尔族入侵西班牙时，由波斯人传播到西方。据说拜占庭帝国也对国际象棋传入欧洲起到过重要作用。

国际象棋起源于印度，但后来成为代表西方的游戏。仔细观察国际象棋的玩法，可以看到游戏开始之前，所有的棋子被放在棋盘方格之内各自固定的位置上。双方各执16个棋子，在横纵均为8格的棋盘上轮流走动自己的棋子，吃掉对方的棋子。在国际象棋中我们要注意的是，棋子分王、后、车、象、马、兵，形成身份体系，而且规定棋子有各自的移动路线。换句话说，所有的棋子只能在规定的

Chess and Go

Games are a media reflecting the characteristics of a culture. The origin of chess is found in the game "*chaturanga*" which appeared in India in ca. A.D. 600. *Chaturanga* was introduced to Persia in ca. A.D. 625. Chess was spread to the West by Persians when Moors invaded Spain in ca. A.D. 711. The Byzantine Empire is also said to have played an important role in the spread of chess in Europe.

Although chess originated in India, it became one of the most popular games in the West. The game is played on a square board with squares arranged in an 8 x 8 square where all the pieces are placed in set positions. Each player has 16 pieces and moves his/her pieces with a goal of checkmating the opponent's king. There is a hierarchy among pieces, such as king, queen, knight, bishop, rook, and pawn; and each piece should move following the fixed pattern. That is, all the pieces are supposed to move according to the hierarchy and geometrically designed paths.

等级体系中，按照规定的路线移动。

围棋的起源有几种说法，其中最有说服力的版本是，围棋约在公元前2300年由中国上古帝王“尧”为教育两个儿子所创。

国际象棋或象棋中有马和象，因为它们是象征游牧社会战争的游戏。围棋则是象征农业社会的游戏，扩张农田，深耕细作。围棋的玩法，简单地说，是在画有方格的棋盘上，围出更多空间的一方赢。对弈的二人分执白子和黑子，在两条线的交叉点上布子，构筑自己的领域。如果一方的棋子围住对方的棋子，就能吃掉里面的棋子，拥有空出来的空间（图16）。在围棋的规则中，重要的不是吃掉对方多少子，而是在棋盘上占据更多的空间。

国际象棋的棋子有着阶级体系，但围棋的棋子只有黑白之分，在对弈双方的

There are many theories about the origin of Go. The most convincing story is that the first Chinese emperor, Yo, created this game to educate his two sons in ca. 2300 B.C.

While chess represents the wars of nomads who used horses and elephants in their wars, Go symbolizes the agrarian society in which fields are cultivated and expanded. In the game of Go, the player who occupies the largest vacant spots on the grid board wins. Two players take turns to place their color of stone – black or white – at the intersections of the grids on the board and build their territory. If one player's stones enclose the opponent's stone (Fig.16), he/she can remove that stone and empty that space. The winner is not the player who takes more of the opponent's stones, but the player who has more void spots on the board.

Whereas the pieces of chess are classified according to hierarchy, the stones of

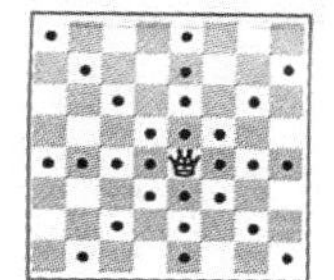
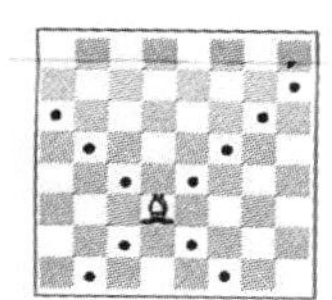
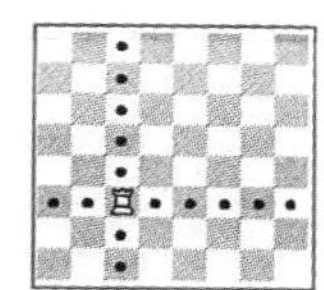
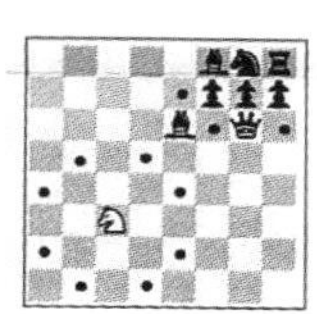
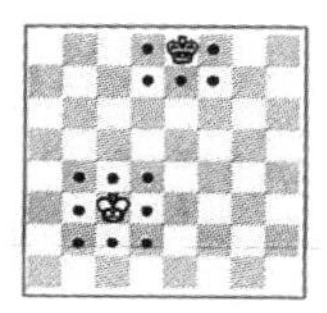
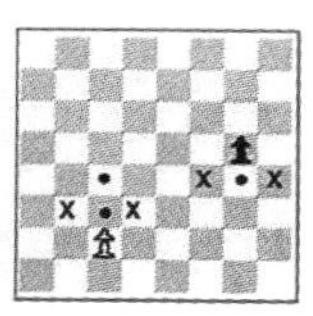

图13
国际象棋中的棋子与移动规则:
皇后，主教（象），车（城堡），骑士（马），
国王，兵（卒）

Fig. 13
Pieces in Chess and the their moving patterns
(queen, bishop, rook, knight, king, and pawn)

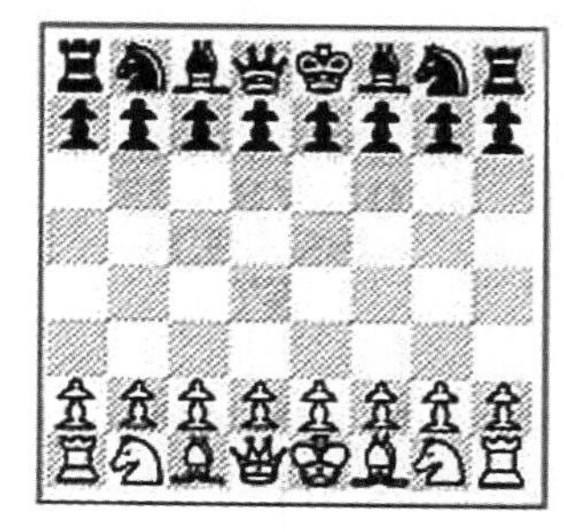

图14
开局

Fig. 14
The initial positions of piece

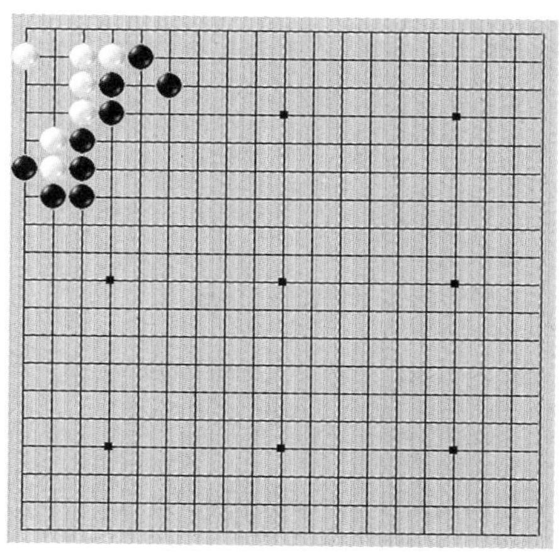

图15
围棋的棋盘

Fig. 15
Go board

阵营中棋子之间没有高低上下之分。每一个棋子地位平等，棋子之间的关系仅仅由它与其他棋子所处位置的相对重要性来决定。比如，图16中，如果白子被黑子围住，白子就会死掉，这时，白子所占位置要空出来，变成黑子的领域。

国际象棋和围棋的不同点可以归纳为四条，这些不同点在两种建筑文化中也有体现。

第一，游戏的规则和构成不同。国际象棋是以吃掉对方棋子、擒获对方的国王为成败的较量，围棋则是由占有更多空间的一方胜出的游戏。

西方建筑大部分是气势压倒外部空间的建筑。金字塔、欧洲城市广场边常见的教堂等建筑，并非包容外部环境，而是压倒外部环境的建筑。站在米兰大教堂或罗马圆形竞技场前，大部分人会被建筑震慑住，产生敬畏感。

Go are divided by color – black and white – and there is no hierarchy at all. All the stones are equal and their destiny is decided only by their relative location. For example, Figure 16 shows that the white stone is removed when it is surrounded by black stones. The space where the white stone was located is emptied and it belongs to the opponent.

The differences between chess and Go can be summed up in four points, which are also the features found in the architecture of the two cultures:

Firstly, the rules and structure of the games are totally different. While the goal in chess is to destroy the opponent's pieces and to kill the king, the goal of Go is to create void space.

Most Western architectural buildings overwhelm their surroundings. Pyramids and many cathedrals in the plazas in Europe do not embrace, but rather dominate their environments. People standing in front of the Milan Cathedral or the Colosseum in Rome are often awestruck by their domineering existence. On the other hand, Eastern architecture is affiliated with its surroundings by absorbing them as a part of it. When we walk through the old palaces in Korea, such as *Gyeongbok Palace* and *Deoksu Palace*, we feel that the low walls, small garden, or space under the eaves in those places are harmonized and the architecture is closely related to

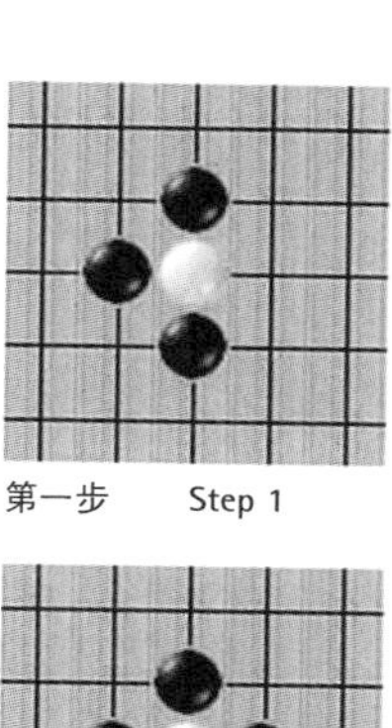

第一步　Step 1

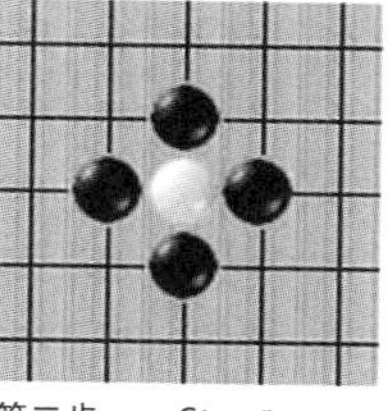

第二步　Step 2

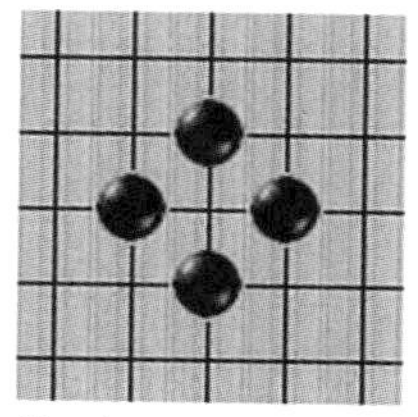

第三步　Step 3

图16
围棋的规则

Fig. 16
The rules of Go

相反，东方建筑通过建筑，将外部环境融为建筑的一部分。走在韩国的景福宫、德寿宫这样的传统建筑前，看着矮矮的院墙、小小的院落以及屋檐下的空间，人们会感到建筑不是孤立的物体，而是与观看的人逐渐融汇，形成密切关系的存在。如果将围棋中每一个棋子当作建筑物，并将它们形成的空间看做是庭院的话，不难发现东方建筑的布局和围棋棋面的相似性（图32，图33）。

第二，国际象棋中，棋子从一开始便具有规定的权力秩序。每个棋子只能按照各自的路线移动。相反，围棋的权利秩序由彼此相关联的、各棋子的相对位置决定。

西方建筑史是不断创造新的建筑样式并反复这种样式的连续。古希腊时期创造出多立克柱式、爱奥尼柱式、科林斯柱式，并将其反复应用到柱子上。哥特

people, instead of being just one of the objects. If we compare stones of Go to architectural buildings, and void space in the grid to garden, we can easily find the similarity between the architectural plans in the East and Go (Fig.32-Fig.33).

Secondly, chess pieces have a power hierarchy from the beginning, and each piece moves according to its own designated routes. In the Go game, however, the power hierarchy is decided by interrelated relative positions during the game.

Western architectural history is filled with successive trials of creating new architectural patterns and their repeated use. The Greeks invented Doric, Ionic, and Corinthian orders and used them repeatedly; and Gothic architecture invented a structural pattern for vaults that widely used flying buttresses.

Western culture sets rules, such as orders, and creates space by repeating those

时期创造出利用飞扶壁支撑的拱券结构，并反复应用。西方文化也是制定样式规则，并通过反复应用规则制造空间。这类似于国际象棋中每个棋子拥有各自的规则和秩序。喜欢制定样式或规则来规定，是西方文化的特点。

而在东方建筑样式中，梁柱结构和样式数千年来始终没有变化。只是，所有建筑根据其所处自然环境来布置，与大自然相协调，形成有机而相对的空间。当然，东方建筑也存在风水等看不见的规则，但就连风水地理关注的焦点也是山、水和人的相对关系。可见，东方建筑比西方建筑更重视相对的关系而不是秩序。

orders, just as pieces in chess follow rules and hierarchy. It seems that one of the characteristics of Western culture is to make regulations with patterns or rules.

For thousands of years, Eastern culture maintained an architectural style that has a structure consisting of wooden columns and beams. All the buildings, created an organic and relative space that is in harmony with nature by responding respectively to the conditions of the land. Of course, there have been invisible regulations, such as *Feng-shui*, but even *Feng-shui* focuses on the relative relationships among water, mountain, and people. In other words, Eastern architecture has put more emphasis on relative relationships than on the orders.

第三，国际象棋的移动规则是棋子在8×8的棋盘上移动。用照相机连续拍下来，应该是复杂的几何图案重叠在一起的图像。而围棋的棋子，落下棋子后不能移动，整体走势是有机增长和不断扩展。

图17是用几何学方式分析菲列波·伯鲁涅列斯基（1377–1446）设计的佛罗伦萨米兰大教堂的穹顶结构图。不只是这张图，从西方建筑大部分建筑图中都能看出，西方建筑是几何图形的组合，沿着几何图形的线条修建墙壁或穹顶，并由此形成空间。

如果将国际象棋棋子在棋盘上的移动路线画下来，会得到类似于图18的几何分析图一样的图案。

几何形、纹样，加上重复运用，经过时间的层叠，西方建筑空间以这样的方式

Thirdly, in terms of moving pattern, chess seems to make repeating patterns on the 8 x 8 squares. It we take a long exposure by opening the iris of a camera lens for a long time, there will be overlapped lines that have complicated geometric patterns. In the Go game, however, the stones cannot change their positions once they are placed, but the overall pattern shows an organically growing and expanding feature.

Figure 17 is a geometric analysis of the dome of the Cathedral of Florence designed by Filippo Brunelleschi. Most architectural plans of Western architecture show repeated geometric patterns. The walls and domes are made according to those lines, creating void space as a byproduct.

If we draw lines following the routes of the pieces of chess, we will get a picture that looks just like the geometric analysis of Figure 17. Void space in Western

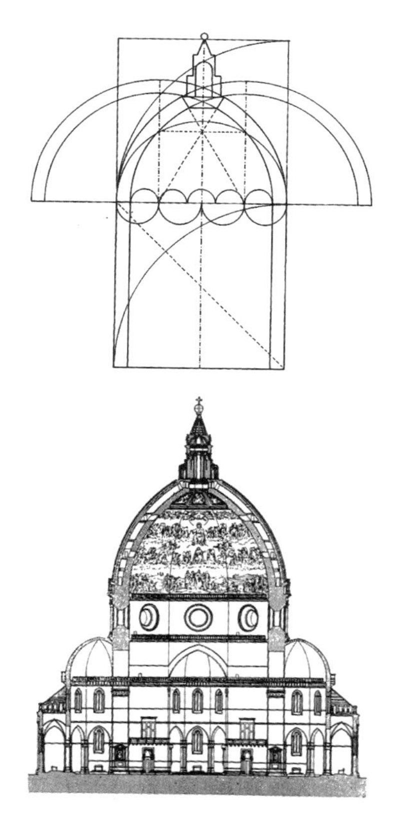

图17
佛罗伦萨大教堂穹顶，
伯鲁涅列斯基1420年设计

Fig. 17
Filippo Brunelleschi.
The dome of the *Cathedral of Florence*, 1420

被设计出来。反观东方建筑，没有固定的规律或重复的样式，像藤蔓般自然扩张。

第四，国际象棋中棋子被放在用线画成的方格内，围棋棋子则放在两条线的交点。

一般来说，西方建筑用厚重的墙壁划分空间，是以墙为中心的建筑；东方建筑则是以柱子为中心的建筑。如图18，西方建筑竖起墙壁，使用由墙壁围合而成的房间。东方建筑竖起立柱，在柱上放置屋檐，这样就形成建筑空间。围棋中所说的“目”，即是棋子连线的内侧，在建筑上就体现为在柱顶盖上屋檐，形成空间。屋檐下的空间有时用墙围合，有时完全敞开。

正因如此，西方建筑的空间是由分割室内外的墙来清晰界定的，给人以坚固而封闭的感觉；东方建筑的内外空间界线模糊，空间很难界定，呈现出流动性。

architecture has been designed by way of geometry, patterns, repetition, and overlapping by time. However, Eastern architecture shows a pattern of vines that expand without defined rules or repeated patterns.

Fourth, whereas the chessmen are placed inside the square made by grid lines, stones of Go are placed at the intersections of the grid lines.

Generally, while Western architecture is wall-oriented architecture, where the space is defined by walls, Oriental architecture is column-oriented architecture. As seen in Figure 18, whereas people in the West set up walls and use the room surrounded by them, people in the East place columns and put a roof on it to create an architectural space. As the "house" in Go refers to the inner space of connected points of stones, columns placed on some points and a roof constitute a house in Eastern architecture. The space under the roof is sometimes closed by

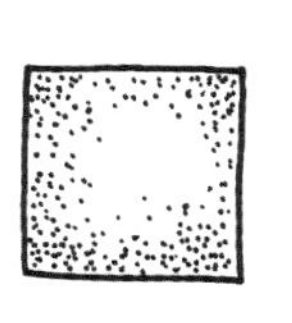
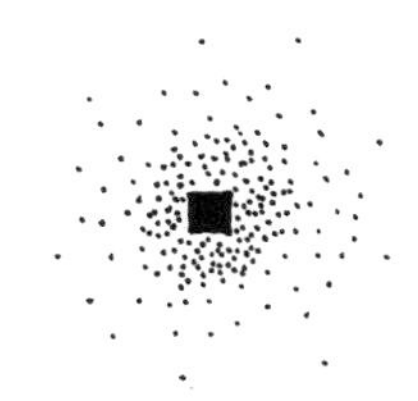

图18
由墙和柱限定的空间

Fig. 18
Territorial claim made by walls and columns

walls, and sometimes totally opened.

Due to this reason, the void in the West is a space whose border between the exterior and the interior is clearly divided by walls, evoking rigid and closed feelings; while Eastern architecture shows an ambiguous boundary between the exterior and the interior and it is hard to define the shape of its void space.

西方的几何性建筑空间

像毕达格拉斯这样的希腊哲学家相信宇宙是按照数学法则运转的，这种对数学根深蒂固的信奉，在建筑、音乐、绘画中也有所体现。在这一概念的基础上，西方建筑运用几何和数学理论设计而成，不断发展。大约在公元前1世纪，罗马的维特鲁威（约前80-约前15）在《建筑十书》中从理论上强调了比例在建筑中的重要性。

若要寻找将形而上学的规则反映到建筑中的倾向之根源，我们可以追溯到古埃及。最近发表的研究成果显示，金字塔群的布局与猎户星座一致。古埃及人观察星象的移动，找出了天体法则，而上天的法则影响到埃及人规划布局巨大的建筑物——金字塔。或许形而上的宇宙规律与形而下的建筑没有任何关联，但西方

Geometric Void Space in the West

Some Greek philosophers, including Pythagoras, believed that mathematical order rules the universe. The deep-rooted belief on mathematics in the West has been applied to such fields as architecture, music, and art. Western architecture has continuously developed with geometric and mathematical designs. Vitruvius theoretically stressed the importance of proportion in architecture in his book *De Architecutra* (*The Ten Books on Architecture*) in ca. 100 B.C.

The origin of the tendency to apply metaphysical rules to architecture can be traced back to Egypt. Recent research shows that the arrangement of the pyramids in Gizeh corresponds to the constellation of Orion. Egyptians figured out the rules of the heavenly body by watching the movement of stars in the sky, and the rules in the sky made an influence in planning the arrangement of the huge architectural building of pyramids. It may seem that there is no relationship between

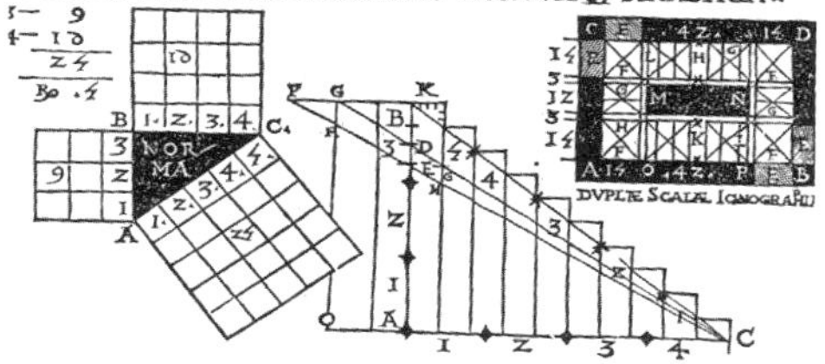

图19
切萨雷·切萨里亚诺1521年翻译的维特鲁威《建筑十书》中的勾股定律

Fig. 19
Pythagoras triangle in *De Architectura*, translated by Cesare Cesariano with his commentary, 1521.

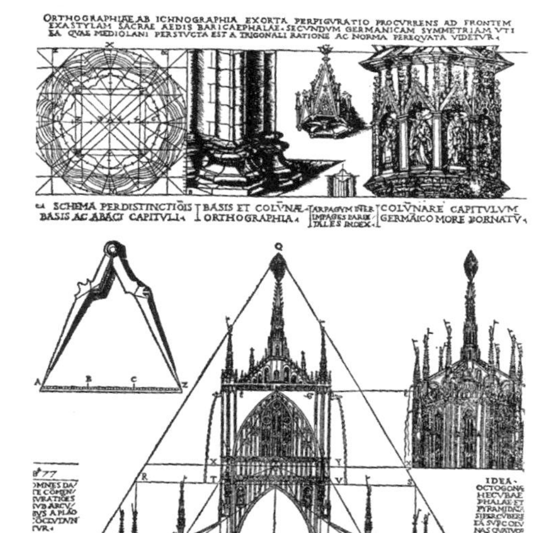

图20
米兰大教堂剖面图
《建筑十书》第一卷第二章

Fig. 20
Illustration of a section of Milan Cathedral in *De Architectura*: Vol. 1. Ch. 2, translated by Cesare Cesariano with his commentary, 1521.

文明的特点就是想从推动宇宙运转的绝对规则中寻找永恒的价值。因此，从西方文明的起源——古埃及开始，西方人便努力把物理或数学的规则反映到建筑中。

将以上分析的东、西方两种文化的特点，以及两者关系用游戏和建筑简单概括下来，如右表所示。

the metaphysical heavenly rules and physical architecture, but the Westerners who searched for an everlasting value in the absolute rule that moves the universe tried to apply physical or mathematical rules to architecture from the Egyptian period, the beginning of civilization.

The basic characteristics of Eastern and Western cultures and their relationships with games and architecture can be summarized as follows.

西方 West		东方 East	
国际象棋 Chess	西方建筑 Western Architecture	围棋 Go	东方建筑 Eastern Architecture
吃掉对方 Goal: killing opponent's king	压倒外部空间 Buildings overwhelm the surroundings	用围棋形成空间 Goal: creating void space	用建筑包容外部空间 Buildings embrace surroundings
绝对的等级 Absolute Class	建筑平面 有绝对的样式体系 Architectural plan has an absolute pattern system	相对关系 决定等级 Class is decided by relative relationship	与地形相呼应 的建筑平面 Architectural plan relatively corresponds to the land
几何形的运动路线 Geometric Route	几何形的空间 Geometric Space	有机的扩展方式 Organically Growing Pattern	建筑布置遵循有机的 扩展方式 Plan shows organically growing patterns
棋子放在方格内 Chessmen are placed inside the grid	墙壁的建筑 Architecture of Wall	棋子放在两条线 的交点上 Stones are placed at the intersections of the grid	柱子的建筑 Architecture of Column

西方建筑空间的数学演化

西方建筑的核心是数学，这种特点在宗教空间中尤为明显。作为接近神、接近真理的手段，西方人选择了具有逻辑性的数学。可以推测，西方人的这种思想很自然地表现在建筑上，形成具有几何学性质的空间。西方的建筑空间基于几何学，仔细观察不难发现其内部也在不断发展。

公元118-128年建造的罗马万神庙，平面图和剖面图上圆形的空间，可以说反映了最基本的欧几里得几何学（图21）。

君士坦丁大帝将罗马帝国的首都从罗马迁到君士坦丁堡，即今天的伊斯坦布尔后，东罗马帝国于公元532-537年在伊斯坦布尔建造了圣索菲亚大教堂。这时开始出现更为复杂的几何图形——与万神庙简单的圆形空间不同，圣索菲亚

The Mathematical Evolution of Void Space in the West

Mathematics is at the heart of Western architecture. It is especially clearly seen in religious space. It is assumed that the Westerners' longing to get closer to God and their choice of logical mathematics as a method to reach the truth naturally created geometric space. The close examination of void space in the West reveals the fact that it has evolved internally.

For example, the plan and section of the *Pantheon* in Rome, which was built in A.D. 118-128, show a void space that looks like a circle (Fig.21). This space was created according to the most basic Euclid geometry.

Roman Empire built *Hagia Sophia* in Istanbul in A.D. 532-537 after the emperor Constantine moved the capital of the Roman Empire to Constantinople (presently Istanbul), and the building shapes based on more complicated geometry appeared

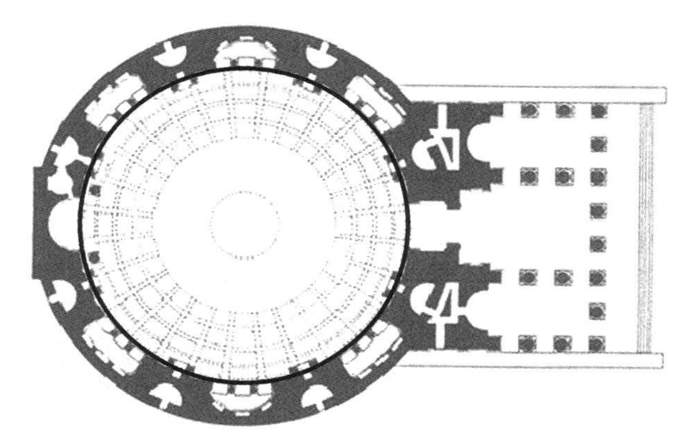

图21
万神庙平面图

Fig. 21
Floor plan of *Pantheon*

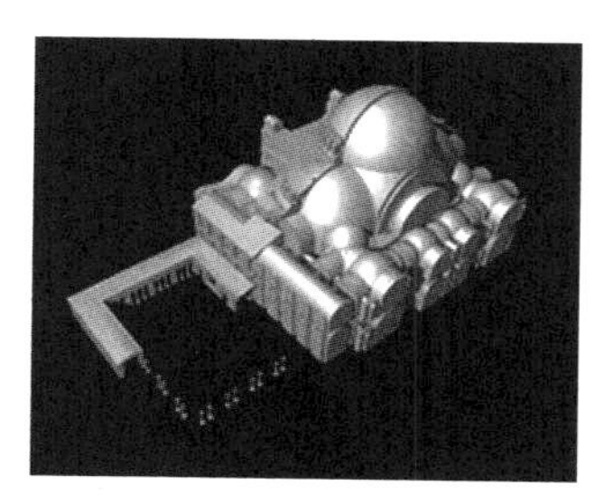

图22
圣索菲亚大教堂的空间

Fig. 22
The volume of void space of *Hagia Sophia*

大教堂将不同尺度的拱顶重叠起来。

如图23，中间有半径为A的三个圆重叠在一起，其周围又有半径为a的圆。半径A和a的比例是3:1。“3”在基督教文化中具有很特别的意义。在希伯来文化中，基于三位一体说，“3”是完整的数字。在西方音乐中使用三度和音也出于同样的原因。圣索菲亚大教堂的主要空间之所以将三个圆重叠起来，估计也是相同的理由。数字“3”还出现在大圆顶和小圆顶的半径比例上，也是3:1。大圆顶的外侧形成走廊，上有小圆顶，上下各有四个。算起来三个大圆顶加上四个小圆顶，圆顶共有3+4=7个。而数字“7”，在基督教文化中被认为是上帝赐予的数字。

观察剖面，所有圆顶上面又有一个圆顶，因此圣索菲亚大教堂的数字有1，3，7。这些数字从基督教意义上看都是神圣的数字。

from this time. Unlike Pantheon which has a simple round shape, the Hagia Sophia shows overlapped vault space in a different scale.

Figure 23 shows three overlapped circles with a radius of A and some circles with a radius of a. The ratio between A and a is 3:1. The number 3 has a very special meaning in Christian culture. According to the doctrine of Trinity, the number 3 represents a perfect number in Hebrew culture. That's why the major chord in Western music is the third chord. The three overlapped circles shape dome space in Hagia Sophia void space seem to be made from the same reason. The number 3 is also shown in the 3:1 ratio between the radii. There are eight domes, four of each at the north and south of the main dome, creating an equation of 3+4=7. The number 7 is known to be a number given by God.

The number "1" is another important number as shown by the fact that the plan

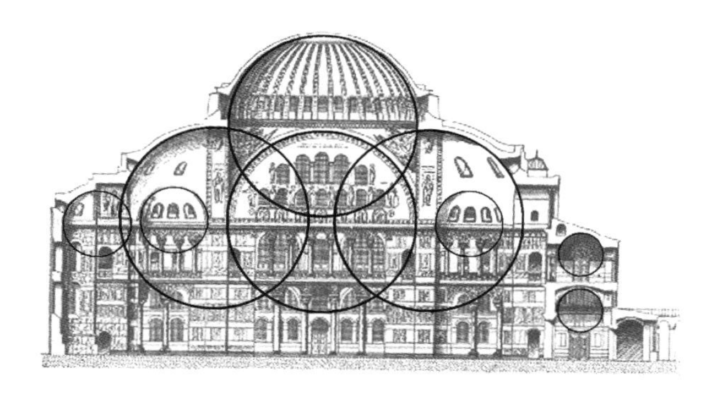

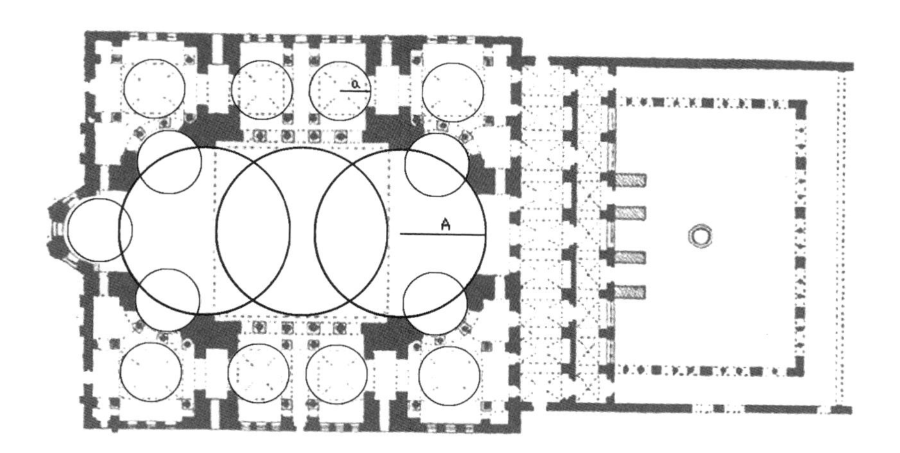

图23
圣索菲亚大教堂的平面与剖面

Fig. 23
Section and plan of *Hagia Sophia*

反复使用同样形状、不同大小的穹顶的方式与数学中的分形理论类似（图24）。分形是用简单的规则制造复杂的形状的方法，人们用混沌理论来解释大自然中不规则的现象时使用的就是这种方法。

西方建筑发展基于数学理论方面的进化，估计与君士坦丁堡的地理位置有关。与罗马相比，君士坦丁堡靠近希腊，而希腊的数学要比罗马更为先进。随着罗马帝国的衰败，很多希腊学者来到东罗马帝国，数学的发展在文化中得到全面的体现。

无论从数学理论上还是建筑意义上看，圣索菲亚大教堂都是如此完美，给当时还是游牧民族、除了搭建移动式帐篷，还没有任何建筑能力的伊斯兰文化，带来了巨大冲击。此后，伊斯兰人效仿圣索菲亚大教堂，发展了自己的宗教建筑——

shows one additional dome on top of every dome. The numbers "1," "3," "7" seen in the *Hagia Sophia* represent the holy numbers in the Bible. The way of repeating the same-shaped dome with different scales is reminiscent of the fractal theory in mathematics (Fig.24). The fractal theory explains a complicated shape with a simple rule, and it is used to explain irregular phenomena of nature by applying the chaos theory.

The architectural evolution based on mathematics in the West seems to have been influenced by the geographical location of Constantinople, which was closer to Greece than Rome. Mathematics was much more developed in Greece than in Rome at that time. After the fall of Greece, many Greek scholars moved to the Eastern Roman Empire and more developed mathematics appeared in overall culture.

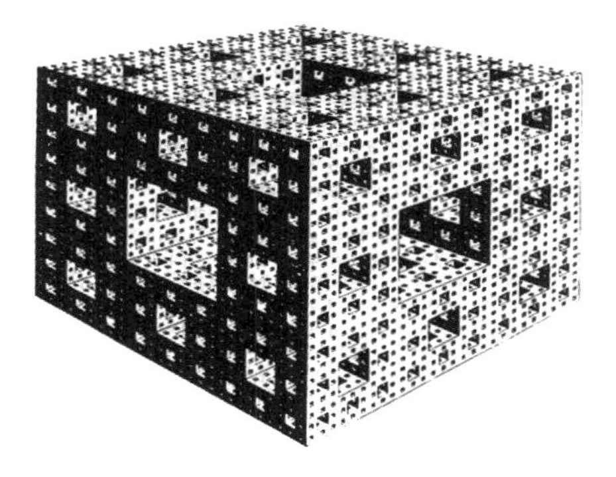

图24
分形理论

Fig. 24
Fractal theory images

清真寺。在数学方面有很高造诣的伊斯兰人在已开始应用分形理论的圣索菲亚大教堂的基础上发展出更为复杂的形式，后来还把阿拉伯式图案应用到建筑中。

公元476年，西罗马帝国灭亡后，欧洲没有实力抗衡处于鼎盛时期的伊斯兰帝国。相反，伊斯兰帝国逐渐把领土扩张到西班牙和北非。同时，高度复杂的几何学形态的伊斯兰建筑对欧洲产生了影响，西方建筑文化的数学特征越来越浓厚。这一时期，在西班牙出现了一种“杂合”的混合建筑样式，阿尔汗布拉宫就是极具代表性的例子。

随着伊斯兰－欧洲杂合式文化的出现和公元1453年东罗马帝国灭亡后，希腊学者大举迁移到欧洲，在欧洲出现了文化复兴建筑样式。伯拉孟特（1444-1514）设计的圣彼得大教堂是文艺复兴时期的巅峰之作，表现出远比圣索菲亚

The mathematically and architecturally beautiful *Hagia Sophia* was a great shock to Islamic people who had no architectural knowledge but to make a movable tent as nomadic tribes. Later, Muslims developed all their temples based on the *Hagia Sophia*. Since Muslims were good at mathematics, they developed the *Hagia Sophia* made by a fractal theory into a more complex form, and also applied patterns such as arabesque to architecture later.

Since the Western Rome Empire was destroyed in A.D. 476, Europe was not well organized enough to fight against the flourishing Islamic empire. The Islam expanded its territory to Spain and North Africa, and as highly complicated and geometric Islamic architecture affected European architecture, the mathematic characteristic of Western architecture was reinforced. Spain created a hybrid architectural style of which the representative example is the *Alhambra Palace*.

大教堂先进发达的数学逻辑的形式。

圣索菲亚大教堂穹顶布局较随意，伯拉孟特则按更为有序的规则布置穹顶。如图25，较小的穹顶位于大穹顶十字中心线的交点上。半径之比——A∶a和B∶b之间的比例大约为0.6。从这里我们可以看出更为先进的分形体系。文艺复兴时期的莱昂·巴蒂斯塔·阿尔伯蒂（1404–1472），在自己的著作中也强调了数学比例在建筑上的重要性。

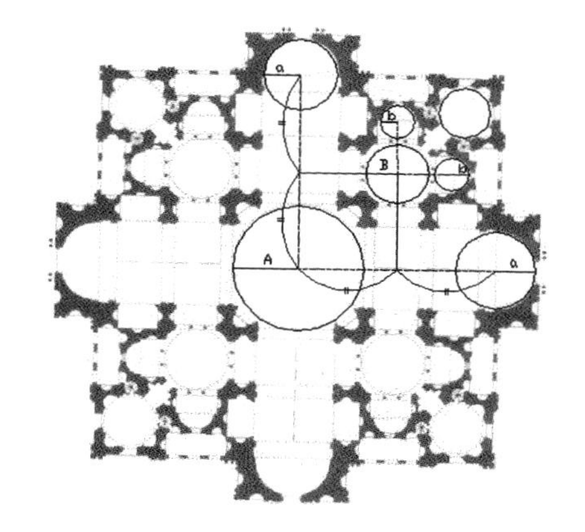

图25
圣彼得大教堂的平面图分析

Fig. 25
Analysis of a plan of *St. Peter's Basilica*

Based on the hybrid culture of Islam and Europe, the Greek scholars, who immigrated to Europe when the Eastern Roman Empire collapsed in 1453, created a Renaissance architectural style in Europe. *St. Peter's Basilica* designed by Bramante (1444-1514) is a highlight of Renaissance culture at that time, and it showed a more developed mathematical form compared to the *Hagia Sophia*.

While the surrounding domes were randomly arranged in the *Hagia Sophia*, the St. Peter's Basilica was made according to more organized rules. As seen in Figure 25, secondary small domes were placed at the intersections of the center lines of the first domes. The ratio between A:a and B:b is about 0.6. It is a more developed fractal system. Besides, an Italian architect Leon Battista Alberti also emphasized the importance of mathematical proportion in architecture in his book.

西方人在解决建筑空间的问题时，总是从几何学角度出发。因此，若找不到简单的解决方法，便会选择更为复杂的数学方法。例如，圣卡罗教堂的建筑师博罗米尼（1599–1667），看到由于基地条件限制无法修建传统的圆形穹顶时，便设计了有两个中心点的椭圆形空间（图26）。

继博罗米尼之后，瓜里诺·瓜里尼（1624–1683）更是触及基于数学理论的空间的极致。瓜里尼的圣辛多纳教堂（图27）复杂的空间布局看上去像一座清真寺，穹顶的平面看上去也像伊斯兰特有的藤蔓式图案。

建筑师、古典学家、数学家、天文学家、设计伦敦圣保罗大教堂的克里斯托弗·雷恩爵士（1632–1723），在他的著作《根源》中说道：

美有两种：自然美和习惯美。自然是由各种相同的或相似的几何形组成

The problems in the architectural space were solved by the geometrical approach in the West, and when there was no simple solution, a more complicated mathematical method was adopted. For example, Francesco Borromini (1599-1667) created void space of oval shape which has two center points when it was not feasible to design a traditional circle-shaped dome in the given land (Fig.26).

Following him, Guarino Guarini (1624-1683) highlighted mathematical void space (Fig.27). The complicated patterns of his *Cappela della SS Sindone* make it look like an Islamic temple. The plan of the dome is similar to that of Islamic arabesque. Sir Christopher Wren (1632-1723), an architect, classicist, mathematician, and astronomer who designed *St. Paul's Cathedral* in London wrote in his book *Parentalia* as follows:

"There are two causes of beauty-natural and customary. Natural is from geometry

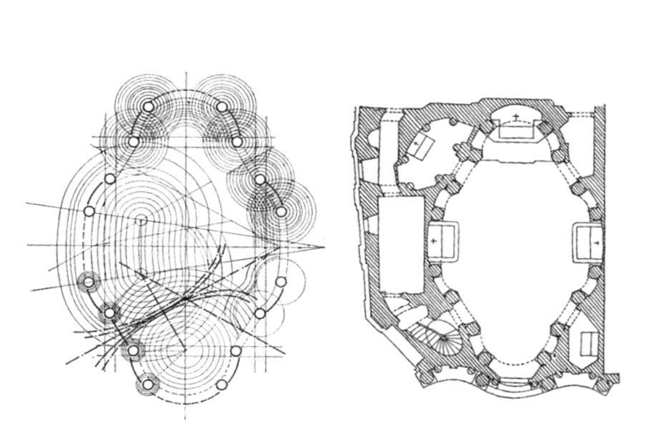

图26
博罗米尼自1634年开始设计的罗马圣卡罗教堂的几何分析，出现了极为复杂的几何形。

Fig. 26
Geometric analysis of *San Carlo alle Quattro Fontane* in Rome designed by Francesco Borromini from 1634. More complex geometry appeared.

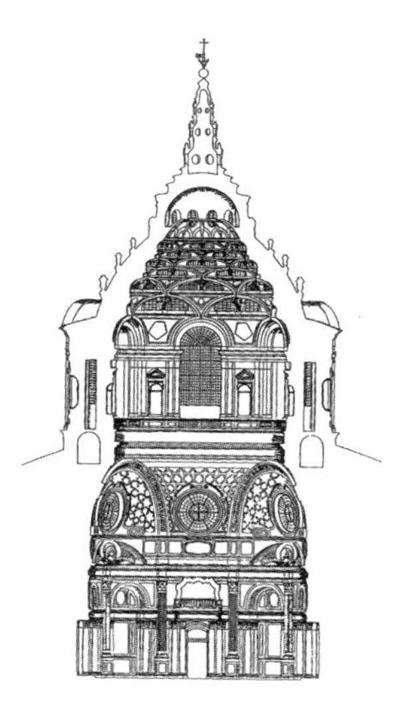

图27
瓜里诺·瓜里尼自1667年开始设计的都灵圣辛多纳教堂

Fig. 27
Cappella della SS Sindone in Turin designed by Guarino Guarini from 1667.

的……几何形比不规则形状美：正方形和圆形最美，接下来是平行四边形和椭圆形。直线只有在两种情况下才能产生美感：一是垂直，二是平行。这既是源于自然，也是必需的，没有比垂直更坚定的形式了。

克里斯托弗·雷恩爵士的这段话充分表明，西方建筑家通过数学几何追求绝对的美。在现代建筑中，格雷戈·林恩（1964- ）或彼得·艾森曼（1932- ）等人的建筑猛一看起来很不规则，但实际上，他们比密斯或柯布西埃更多地继承了西方建筑空间将数学理论结合到建筑设计的传统，只不过是利用电脑绘制出更为复杂的形式（图28，图29）。

consisting in uniformity, that is equality and proportion......Geometric figures are naturally more beautiful than irregular ones: the square, the circle are most beautiful, next the parallelogram and the oval. There are only two beautiful positions of straight lines, perpendicular and horizontal; this is from Nature and consequently necessity, no other than upright being firm."

This passage clearly shows the fact that Western architects searched for absolute beauty through mathematical geometry. The designs of modern architects such as Greg Lynn or Peter Eisenman look very irregular at a glance (Fig.28, 29). The fact is that they look complicated just because they adopted the means of computer, and they are nevertheless the works that succeeded the genealogy of traditional Western architectural space that tried to apply mathematical rules to architectural design more than did those of Mies or Corbusier.

图28
格雷戈·林恩的设计

Fig. 28
Designed by Greg Lynn

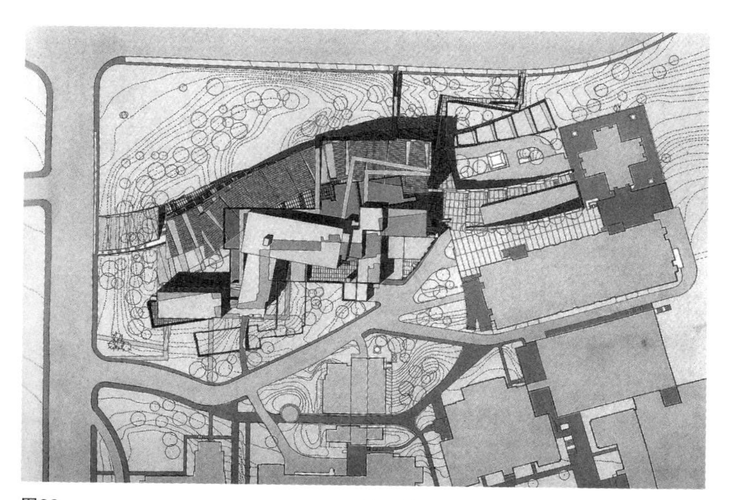

图29
彼得·艾森曼的设计

Fig. 29
Designed by Peter Eisenman

东方建筑——柱网系统与模糊的分界

与西方建筑不同，东方建筑没有可资了解建筑体系发展过程的样式变化。从结构的角度看，东方建筑是基于柱网体系的。

从图30可以看出，不管什么年代的东方建筑，都是在网格的交叉点上立起柱子，在两个柱子顶上搭屋面。这种柱网系统与围棋的体系非常相似。东方建筑的柱子位于交叉点上。同样，围棋的棋子也是放在棋盘上横竖两条线的交叉点上。而且每个柱子和棋子都能够在基地或棋盘上划定自己的领域。图30是日本建筑的柱网体系，让人联想到围棋的棋盘。如果把棋盘看做是为建筑师提供的土地，可以发现围棋棋盘上的棋子布局，与东方建筑表现出的可以有机增生的平面形态非常相似。图31中若把围棋的黑子看做建筑，白子看做树木、石头、水等自

Eastern Architecture with a Column System and Ambiguous Boundary

Unlike Western architecture, there is no change in architectural orders that shows the evolution of architectural style in the East. Structurally speaking, Eastern architecture is based on grids and columns.

Figure 30 shows the fact that all the buildings in the East have been made by placing columns of even numbers on the intersections of grid and a roof on top of the columns throughout history. The system of grid and column is very similar to that of Go. As stones were put on the intersections of grid in Go, the columns were placed on the intersections of grid in Eastern architecture. The role of each column and stone is to claim its own territory. The diagram in Figure 30, a grid system used in Japanese architecture, reminds one of a Go game board. If the Go game board is compared to the land given to an architect, the patterns of the placement of stones are compared to the organically growing pattern shown in

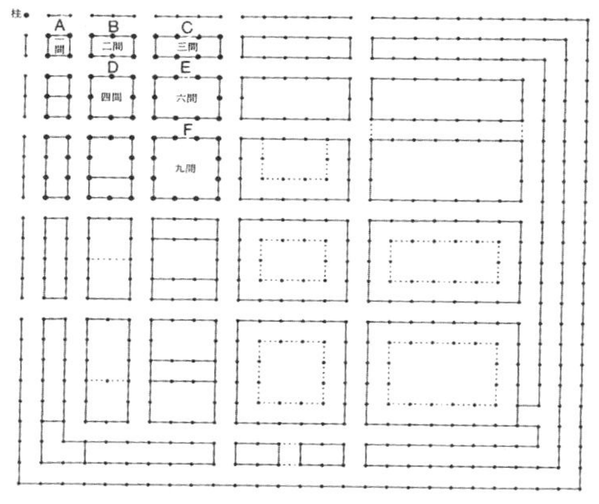

图30
东方建筑中的柱网体系

Fig. 30
Grid and column system in Eastern architecture

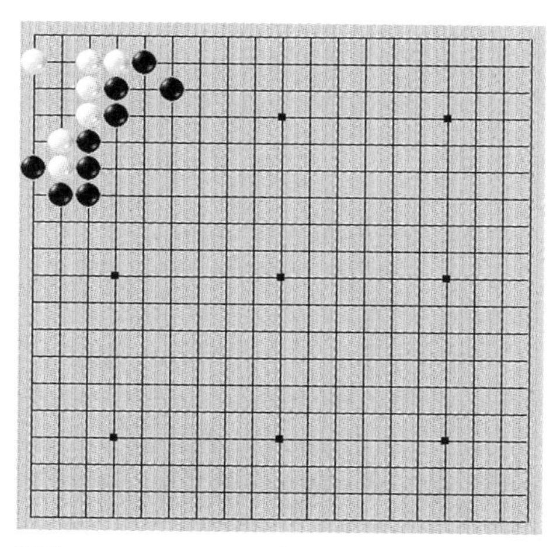

图31
围棋的棋盘

Fig. 31
Go game board

然元素，棋盘上的棋子布局（图32）与图33中桂离宫的平面图极其相似。

东方建筑的基本要素可以归纳为立柱、屋顶、矮墙。像围棋棋子一样，这些要素的价值由它们之间的相对关系决定。正如中国汉字，仅仅改变笔画的相对位置就可以形成新的汉字一样，上述三个建筑要素在所处的自然环境中共同营造出新的价值。在东方，建筑建造的过程中总会形成各式各样的庭院。如果联想到围棋的规则——由所围空间大者胜出，以及老子有关“无”的思想，就能够理解东方建筑中的园林和庭院是如何形成的了。

万物负阴而抱阳。

老子《道德经》第二十八章

the plan of Eastern architecture. The arrangement of stones and the plan of *Katura Villa* look very similar if the black stones are compared to architectural buildings and the white stones to natural elements, such as trees, stones, or water on the site.

The three basic elements of Eastern architecture can be defined as a column, a roof, and a low wall. The value of these elements is decided by relative relationships, like the stones on the Go game board. As Chinese characters are newly created by changing the relative location of basic strokes, these three elements create new meanings by influencing each other under the given natural conditions. There are various types of gardens made when buildings are constructed on the land. The rule of Go which decides the player who creates more void space as winner, and Lao Tzu's "thought on emptiness" will help people understand the way that gardens and yards are made in Eastern architecture.

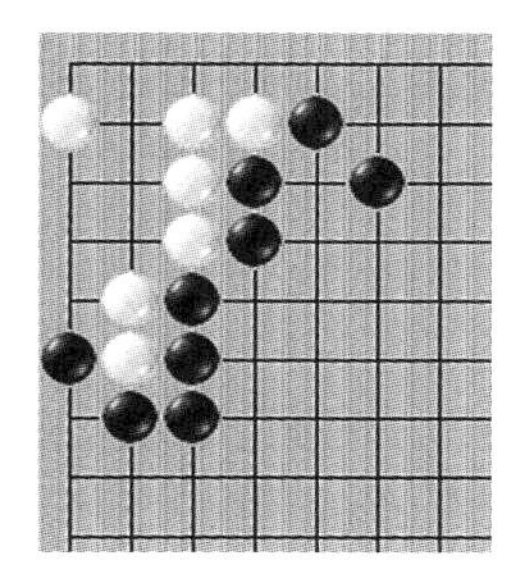

图32 Fig. 32
围棋的棋面 Patterns in Go

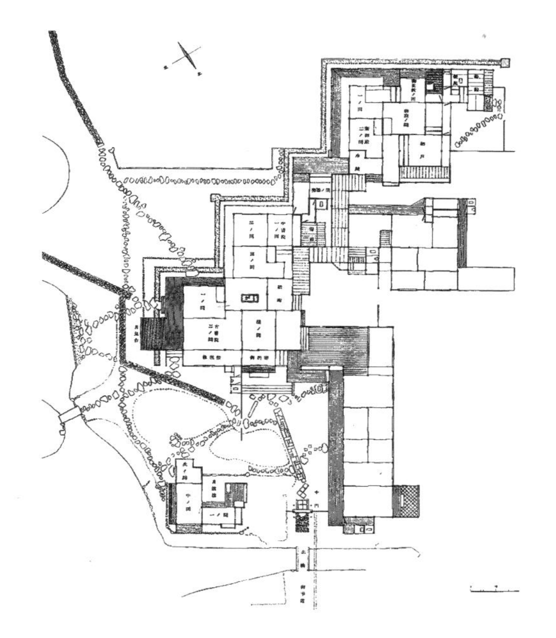

图33
京都桂离宫
（1615—1663年）

Fig. 33
Katura Villa,
1615-1663, Kyoto

"The way to acquire Yang is to contain Yin."

Lao Tzu. *Tao Te Jing*. Ch. 28.

老子将“无”看做是“一种随时可以被填满的状态”。受道家思想影响的枯山水在空地上铺沙以象征大海，放上几块石头象征岛屿，余下都是空白。大海因为风掀起浪涛，但在枯山水中由沙象征的大海是纹丝不动的，它是静止的，也是“永恒”的空间。东方建筑的空间建立在柱网系统之上，蕴含着无限的可能性和永恒。西方在几何性的建筑空间内添置雕塑、彩色玻璃、壁画等具有象征意义的元素，创造宗教性质的空间；在东方，通过留白实现具有宗教意义的空间。

日本京都龙安寺石庭院中有十五块岩石（图34）。这些石块的布局不管从哪个方向看，都只能看到十四块，总有一块被其他石块遮挡。据说，这是为了告诉人们“知足之足常足矣”的道家思想。从中可以看出，在东方建筑中第一人称的认知非常重要，东方建筑在设计时会考虑相对位置的关系。

According to Lao Tzu, a void is "a state that has a 100% possibility of being filled in the future." In a Zen garden, influenced by Zen thought, there are only a few stones symbolizing islands on the sand that represents the sea, and the rest remains as an empty space. The metaphorical sea of Zen garden disturbed by the wind because it is made of sand. It is space where time stops and eternity begins. With the implication of possibility and eternity, the void in Eastern architecture is based on a system consisting of columns and grids. While the religious space is made in the West by filling the geometric void space with materials that have symbolic images, such as sculptures, stained glass, and paintings, the religious space is created by emptying in the East.

There are 15 stones in the dry landscape garden in *Ryoanji* (Fig. 34), and only 14 out of 15 stones are seen from every direction, leaving one stone hidden by other stones and invisible from any direction. It is said that this is to deliver the teach-

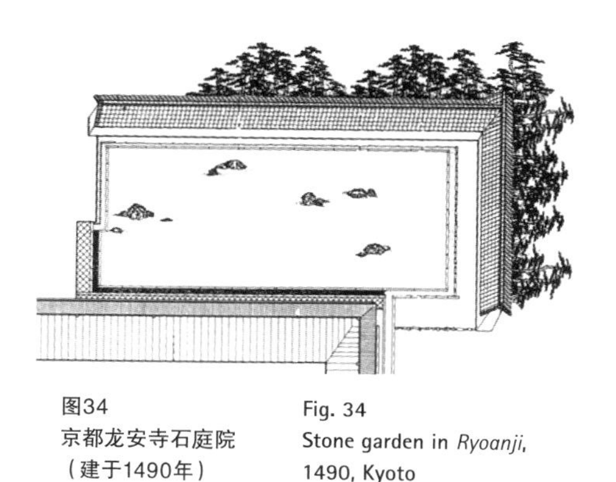

图34
京都龙安寺石庭院
（建于1490年）

Fig. 34
Stone garden in *Ryoanji*, 1490, Kyoto

ing of Taoism which says "Be satisfied even if things are not filled up." It reflects the fact that the perception by the first person is important in Eastern architecture, and its design takes the relationship based on relative positions into consideration.

Whereas void space in the West is clearly divided by geometrically designed walls and roofs, void space in the East has a ambiguous boundary. Lao Tzu said "*The basis of two things is one. It is just that the two things have different names. The secret lies in the united harmony between the two.*" The Eastern philosophy divides *Yin* and *Yang*, but it actually put more emphasis on harmonizing the two.

西方建筑中的空间由遵照几何学方法设计的墙壁和屋顶明确划分，而东方建筑的空间是界线模糊的。关于有和无，老子说，“此两者，同出而异名。同谓之玄，玄之又玄，众妙之门”。东方哲学将世界分为阴和阳，但更重视阴阳合一。

同理，东方建筑的空间追求室内外空间的协调，以及自然与建筑的统一。因此，东方建筑的空间是自由的、流动的空间，而不是刻板的、被限定的空间。用建筑术语来说，就是“开放平面”。

在建筑平面图中，墙用线来标记，柱子用点来标记。墙和柱子所构成的空间的不同正如线和点的区别。如图18，线划分出区域，树立边界，形成空间；而点仅仅暗示模糊的领域。因此，立柱构建出的建筑空间边界模糊，激发出空间的流动感。

For the same reason, void space in Eastern architecture has always pursued the unity of two objects through the harmony between the interior and the exterior or the integration of architecture and nature. Thus, void space in Eastern architecture is flexible and fluid, rather than rigid and defined. In architectural terminology, it is an "open plan."

In the architectural plan, walls are expressed by lines and columns by dots. The spatial difference created by walls and columns is the same as the one between lines and dots. As seen in Figure 18, whereas lines divide an area, build a boundary, and create voids, dots just imply an ambiguous territory. The void created by columns, therefore, creates space that has an ambiguous boundary that evokes a fluid feeling.

东方和西方的绘画

东、西方的绘画作品也很好地体现了各自的文化特点。西方绘画从古埃及开始就信奉“黄金分割律”，画布上的所有元素都根据“黄金分割律”这一数学规律细心布局。与之相反，东方绘画重视“留白”而不是黄金分割律，留白与画作描绘的对象同样重要。这种绘画的理念源出老子的哲学。

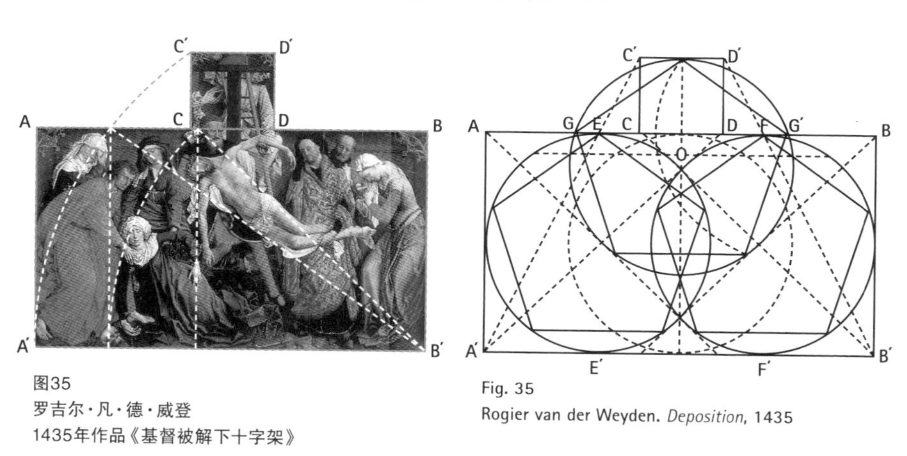

图35
罗吉尔·凡·德·威登
1435年作品《基督被解下十字架》

Fig. 35
Rogier van der Weyden. *Deposition*, 1435

The Paintings of the East and the West

The cultural characteristics of the East and the West are found also in paintings. Golden ratio has been widely used in the West since Egyptian period up to now. Every element on the canvas was carefully designed according to mathematical elements of the golden ratio. In the East, on the other hand, the "beauty of blank space" was counted instead of the Golden ratio. The background margin was as important as the figure itself. This concept seems to originate from Lao Tzu's philosophy.

东方绘画中，图底之间是相辅相成、相互渗透、均衡的关系，与围棋的棋面和东方建筑的平面很相似。如图36，在远景和近景之间，画家没有画中景，而是留下空白。这种表现方式也出现在东方建筑中。东方建筑常使用独立的矮墙，这些矮墙遮蔽了中景，在透视图上产生视觉上的留白，只留下近景庭院和远山，与东方绘画中的留白属同样的妙用。

有时铺沙的庭院，表现出更多的空间（图37）。在东方，设计并不遵从黄金分割律。在西方的建筑和美术作品中黄金分割律无处不在，而在东方的建筑和美术作品中，空间或留白与实体一样受重视。

The complimentary, interpenetrating, and balanced relationship between the figures and background in Eastern paintings is similar to the patterns of Go and the architectural plans. In Figure 36, the painter created margin space instead of a middle view between the near landscape and the far landscape. It is the technique used in Eastern architecture, too. In Eastern architecture, low, free-standing walls, which act as a margin in Oriental paintings, often appear to block the middle landscape to create a visual margin in the perspective, leaving only the front garden and the rear mountain. Sometimes a void sand garden is located to introduce more void to the architecture (Fig.37). The designs in the East do not follow the Golden ratio. While the geometric Golden ratio was the essential principle in the West, void was as an important element as solid figures in architecture and paintings of the East.

图36
15世纪中期日本画僧文清的山水画

Fig. 36
Bunsei. Mid-1400s

图37
龙安寺

Fig. 37
Ryoanji

东西方文化的混杂

Merging of Two Parallel Lines

东西方文化的混杂

Merging of Two Parallel Lines

中国文化传入西方

前面分析了东、西方文化的特点。在这一章，将回顾两种文化是如何交流，进而创造出新的杂合。

欧洲与中国在罗马帝国和汉朝时就通过丝绸之路开始了贸易活动。但是直至16世纪，文化思想领域才开始交流。第一次文化浪潮是从中国传入欧洲。1669年出版了东印度公司特使自1663年起在中国旅行的详细记录，并附有插图。1687年，最重要的介绍孔子哲学的译著首次在巴黎出版。这些举措令17世纪的欧洲接触到全新的、充满异国风情的中国文化。中国文化的引入，令欧洲知识分子崇拜中国，甚至有人因为热爱中国的艺术和科技，希望自己能够出生在中国。

The Inflow of Chinese Culture to the West

The characteristics of the Eastern and Western cultures were examined in the previous chapter. In this chapter, the history of how the two cultures began to exchange and create hybrid value will be studied.

The trade between Europe and China via the Silk Road started since the Roman Empire and Han Dynasty. It was not until 1500s, however, that the cultural or intellectual exchange began. The first cultural influence that spread to Europe came from China. In 1669, a detailed report, with illustrations, made by an ambassador of the East India Company who traveled through China from 1663 to 1657, was published. In 1687, the first important translation of the book on the philosophy of Confucius was published in Paris. By the 17th century, the West was fascinated by new and exotic Chinese culture. There were some European intellectuals who adored China. So impressed were they by the arts and science of China that some

中国文化持续传入欧洲。1757年，钱伯斯（1723-1796）出版了《中国房屋设计》一书。18世纪，欧洲的启蒙主义哲学家认为儒家哲学足以代替基督教。有趣的事，当时的中国利用基督教消除儒家思想中佛教和道教的影响。流传至欧洲的中国文化逐渐地对西方文化产生影响，最早体现在园林艺术——英国的风景式园林中。

Europeans even wished that they would rather have been born in China. Chinese culture was spreading continuously in Europe. In 1757, Chambers published *Design of Chinese Buildings*. In the 18th century, European Enlightenment philosophers found Confucianism as an alternative to replace Christianity. Interestingly, it is said that Christianity was used to eliminate Buddhist and Taoist elements from Confucianism in China at that time.

Chinese culture gradually began to exert an influence on European culture, and the change is first seen in landscape design starting from the Picturesque Movement in England.

中国陶瓷的作用

陶瓷的出口，可以说是中国文化迅速传播到西方的主要原因之一。仔细观察16世纪的西方绘画作品，可以发现当时的欧洲贵族使用沉重的金属餐具。对于当时的西方人来说，轻巧的、绘制着美丽的彩色图案的中国陶瓷带来的惊喜远超出今天人们对“苹果”——尖端科技产品的喜爱。

西方人从中国进口新的餐具——瓷器，乃至今天英文中仍把瓷器叫做“china”。为了从中国进口陶瓷，欧洲人源源不断地将墨西哥白银运至中国，清王朝由此积累了大

图38
清朝的瓷器，瓷器上刻画的中国建筑与生活方式影响到西方的建筑。

Fig. 38
Chinaware of Ching Dynasty
Chinese architecture and lifestyles depicted on porcelain affected Western architecture.

Chinaware in the West

One of the main reasons why Chinese culture spread so rapidly in the West was chinaware. The 16th century paintings in the West reveal that European aristocrats used heavy metal tableware at that time. For them, the light chinaware painted with beautiful colors meant more than high technology.

Since the Westerners first encountered the new type of tableware imported from China, it is still called "china." The West continuously imported chinaware from China, and China made a great fortune and restored the Great Wall of China using Mexican silver it received as payment for chinaware. When civil wars destroyed

量的财富，用以重修长城。但由于农民起义，中国的窑厂遭到破坏，日本趁机取代中国成为陶瓷的主要出口国。

当时，日本人使用印有木刻版画的纸张包装瓷器，这些包装纸上的木刻版画得以在西方广为流传，日后甚至影响到凡·高（1853–1890）的绘画。

古罗马时期，西方人对中国尤其是中国的丝绸十分向往，但由于当时的货物运输只能通过丝绸之路，非常艰辛，因此丝绸成为极少数富裕阶层的特供品，价格堪比黄金。但从15世纪开始，西班牙和荷兰主导的航运业日渐发达，西方人将大量的中国商品运入欧洲，中国文化对西方文化的影响更广泛了。综上所述，瓷器和航运对中国文化传入西方起到重要的作用。

chinaware plants in China, however, Japan secured a bridgehead for the export of chinaware.

Japan exported chinaware wrapped in wrapping paper printed with woodcuts, and upon this opportunity Japanese woodcuts were introduced to the West, exerting an influence on such paintings as Van Gogh's in later days.

Westerners had such a longing for China that the Chinese silk imported to Rome during the Roman period was sold at the price of gold, and since the silk was imported only through the Silk Road, it was only available for the extremely rich people. Since the 15th century, when the distribution of products through seaway was developed under the leadership of Spain and the Netherlands, large quantities of products from China were imported by the West and Chinese culture became more widespread in Western culture.

从几何式园林到风景式园林

从古罗马时期开始，西方的园林设计就是几何式的。西方园林力求在方形的基地上自创一个宇宙。

从中可看出西方人的世界观。对秉持机械性思维的西方人来说，宇宙是遵循数学规则制造出的完美。因此，西方人认为在创造另一个大自然——造园时，也应该追求几何学、数学上的完美。

图40是一幅描绘“结纹花园”的木

From Geometric Garden to Picturesque Garden

Since the Roman period, gardens were designed to have geometric shapes in the West. The Western garden designers seem to have tried to create a self-completing cosmos in square land.

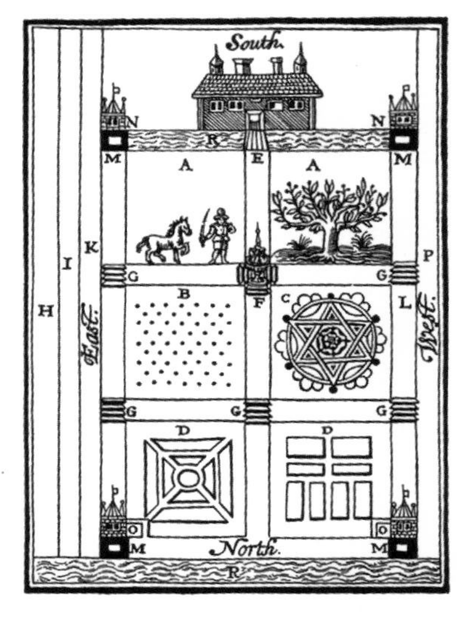

图39
威廉·罗森《新型果园和花园》（1618年）中的木刻版画

Fig. 39
Woodcut from *New Orchard and Garden* written by William Lawson (1618).

The garden design reflects Westerners' world view. From a rather mechanical point of view, Westerners perceived the universe as a perfect creature made according to mathematical rules. Therefore, they believed the garden design which creates another nature also should seek for geometric and mathematical perfection.

Figure 40 is a woodcut showing a design of "Knot Garden." It was excerpted from

刻版画，选自1615年出版的《乡间农场》，作者为马卡姆（1568–1637）。这幅画中的花园构图令人联想到国际象棋棋盘。花园中每块四方形的分隔都有独特的设计，与国际象棋的方形棋格由按规则移动的棋子占据的方式非常相似。

正如图40所示，传统的英国园林都是按照几何图案设计的。但这种"僵硬"的设计在中国文化的影响下开始消亡。后来，瑞典籍美

图40
"结纹花园"实例，马卡姆《乡间农场》（1615年）中的木刻版画

Fig. 40
Design samples of Knot Gardens. The woodcuts are from the *Country Farm* written by Gervase Markham (1615).

Country Farm, written by Gervase Markham and published in 1615. The composition of the garden in this picture is reminiscent of a chessboard. The square-shaped divisions in the garden are filled with self-completing designs, as the squares on the chessboard are occupied by pieces that move according to the rule.

As seen in Figure 40, traditional English gardens were designed with geometric patterns. With the influence of Chinese culture, however, the rigid design began to disappear. According to Osvald Sirén, the late Swedish historian of Chinese art, the Picturesque garden design was inspired by Chinese philosophy and Chinese garden design to some extent. The Chinese philosophy and garden design changed Europeans' mindset toward nature; and they changed the geometric and

术史家、汉学家奥斯瓦尔德·喜龙仁（1879-1966）评价道，风景式园林在一定程度上受到中国哲学和中国园林的影响。显然，中国的哲学和园林设计改变了欧洲人对待大自然的态度，这种变化直接反映在园林设计上，改变了欧洲过去几何性的形式主义园林设计，转换为以认知为导向的风景式园林。

风景式园林设计的杰出代表汉弗莱·雷普顿（1752–1818）说：“是高山还是平原，取决于你是站在哪里看的。”[4] 在设计园林时，他强调观者的认知，因为观察者的位置决定园林中各组成元素之间的关系。在雷普顿的设计图（图41）中可以看出，地平线与牛、人、树木的相互关系随观者的位置变化而变化。这与汉字中“本”和“末”，由于笔画“一”的位置不同，形成完全不同的意思相仿。

老子说，“大直若曲……大方无隅……大象无形”。基于老子的这种哲学，

formal garden design of Europe to perception-oriented design. The most prominent Picturesque garden designer Humphry Repton (1752-1818) said, "A plain a hill, or a hill a plain, according to the point of view from whence each is seen." [3] When he designed a garden, Repton emphasized the perception of the person who is in the garden. He said that the location of a viewer changes the relationships among garden elements. The relationships among cows, men, trees, and the earth are determined according to the viewer's location (Fig.41). It is similar to the case where the different location of '一' stroke changes the whole meaning of Chinese characters such as '**本**' and '**末**.'

Lao Tzu said, "*The greatest straight line looks as if it is a curve... and the greatest rectangle has no corners... The greatest image has no form.*" Based upon this

---

[3] **Humphry Repton,** *Observations on the Theory and Practice of Landscape Gardening.* **(Oxford: Phaidon, 1980), p. 25.**

中国的园林设计中多用曲线，以此体现中国哲学追求的最高境界之一——天人合一。受中国文化的影响，风景式园林造园师们在园内建造中式的楼台亭阁，摒弃直线，引入自然的曲线，在园林内创造出更多的空白空间。正如1739年的设计“斯托花园”，在1753年的改建中，引入更多的曲线，各种边界也变得模糊（图42，图43）。

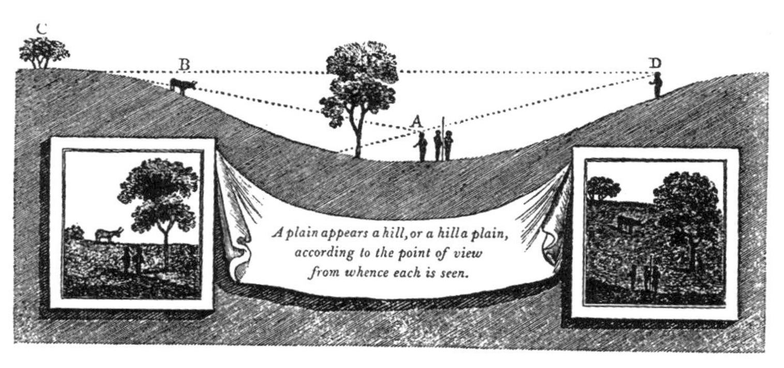

图41
汉弗莱·雷普顿1803年著《造园理论与实践的观察》中的插图

Fig. 41
Illustration from the book, *Observations on the Theory and Practice of Landscape Gardening* written by Humphry Repton (1803)

philosophy, traditional Chinese gardens were designed with curved lines to be united with nature, achieving one of the ultimate goals of Chinese philosophy. Influenced by a Chinese concept, Picturesque garden designers incorporated Chinese pavilions in their gardens, used natural curves instead of geometric design, and created more void space. For example, whereas the original *Stowe garden* designed in 1739 had artificial straight lines and geometric axes, the design made in 1753 shows more curves and ambiguous boundaries (Fig.42, 43).

汉弗莱·雷普顿在修改阿丁罕大道的设计时，通过移除树木获得更多的空间，也是受到东方文化的影响（图44，图45）。有些西方历史学家们认为风景式园林受让·雅克·卢梭（1712-1778）的影响，但卢梭在1758年才正式出版他的第一部著作——《论人类不平等的起源和基础》，因此从时间上看西方园林设计风格的变化在卢梭改变欧洲文化范式之前就已经开始了。相反，应该说卢梭受到因东方文化而改变的欧洲文化范式的影响。

风景式园林重视个人体验与认知。西方传统的几何式园林设计并不关心园林中各种设计元素之间的关系，而是以旁观者的视角为考量。但风景式园林设计师意识到观察者所处的位置，无论是发生水平向还是垂直向的变化都会影响到园林中各组成元素之间的关系，并将这种影响反映到设计中。

In a revised version of design for *A Drive at Attingham*, Humphry Repton created void space by removing trees, showing the influence of Oriental thought, which makes the most of void space (Fig.44, 45). Some Western historians insist that the Picturesque Movement was influenced by Jean Jacques Rousseau (1712-1778). But, since it was 1758 when his first official book *Discourse on the Origin and Basis of Inequality Among Men* was published, it is more likely that the new garden style in the West emerged before Rousseau changed the cultural paradigm in Europe and the new cultural paradigm influenced by Eastern thought affected Rousseau.

Picturesque style is based on people's experiences and perceptions. Traditional geometric gardens were designed from the third point of view, and the relationship between elements inside the garden was not a designer's major concern. But, Picturesque garden designers were aware of how the viewer's horizontal and ver-

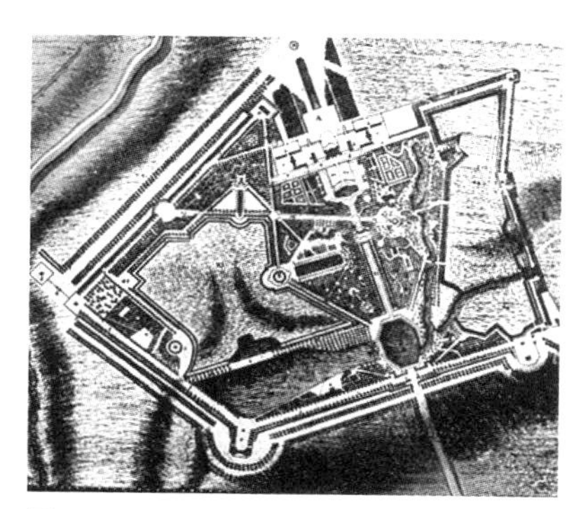

图42
引自莎拉　布里基曼拍摄的
斯托花园平面 (1739年)

Fig. 42
Floor plan of the *Stowe Garden*
published by Sarah Bridgeman (1739)

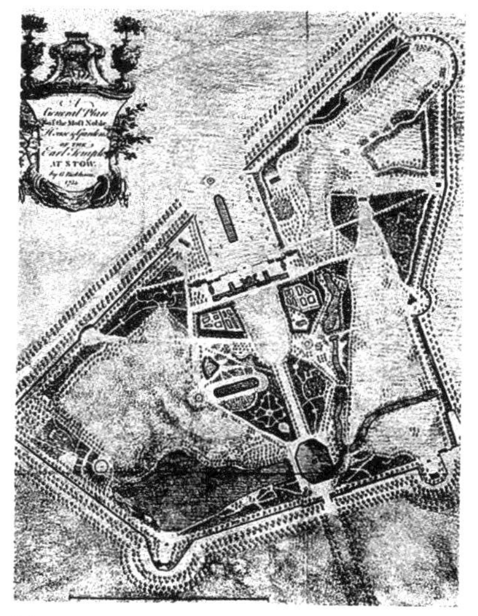

图43
引自彼克曼拍摄的斯托花园平面(1753年)，与1739年的平面相比，包括扩建的花园主路在内的许多园路都从直线形改为自然的曲线形，在建筑前也增加了一大片开敞空间。

Fig. 43
Floor plan of the *Stowe Garden* published by Bickham (1753). Compared to the plan made in 1739, most roads – including the main roads in the extended garden – were modified using curved lines, and a large void space in front of buildings was added in this edition.

人们走在风景式园林中，根据自己在不同位置上所看到的不同景象，在头脑中组成这个园林的整体形象。可以说，在西方园林设计中，相对关系首次占据了重要的位置。可以说在西方文化范式的重点从静态几何学到相对论的转移中，风景式园林设计起到重要的作用。

18世纪风景式园林的设计师强调“自然”的价值，积极地推广“留白”空间的价值，此时空白空间不再是黄金分割律和几何学设计之后剩下的副产品。自18世纪开始，西方文化对空白空间的认识发生了变化，从原先只关注于实体，到开始在虚空空间中发现美。这种转变从19世纪开始在艺术领域中得到表现。

tical location affected the relationship among garden elements, and they reflected that relationship in their designs. People who walked through a picturesque garden experienced the relationships among the elements that change according to their locations. People perceived different images from different locations in the garden, and cognitively constructed the totality of the garden space. The relative relationship became important for the first time in the history of garden design in the West. Picturesque garden design played an important role in Western culture, in that it shifted the focus of paradigm from static geometry to relativity.

As the 18th century picturesque garden designers stressed the "natural" values, they introduced more void in an active way. The void was not a byproduct of geometry and the Golden ratio anymore. The understanding of the void was changed in Western culture since 18th century, and the people started to find the beauty in the void itself instead of focusing only on solid objects. This shift began to appear in fine art since the 19th century.

图44
汉弗莱·雷普顿最初设计的阿丁汉大道，
由一排大树围合。

Fig. 44
A Drive at Attingham, designed by Humphry Repton.
The existing condition shows the drive is enclosed by a row of trees.

图45
汉弗莱·雷普顿修改后的阿丁汉大道设计，
移除了树木形成更开放的空间。

Fig. 45
A Drive at Attingham, designed by Humphry Repton.
His design shows void space created by removing some trees.

东方文化与现代艺术

17世纪，马可·波罗神话般的游记与很多商人的冒险故事传入欧洲，欧洲兴起一股模仿和学习中国的潮流，被称作“中国风”。

“中国风”是持久地盛行于欧洲社会的文化思潮之一，它几乎席卷了与设计相关的各个领域，包括装饰、家具、园林建筑、餐具、挂毯等。

19世纪的西方人开始关注东方绘画作品，其中一个家喻户晓的故事就是凡·高所受的日本浮世绘的影响。尽管凡·高从日本浮世绘中汲取的是色彩的影响，而蒙德里安（1872–1944）、特奥·凡·杜斯伯格（1883–1931）和胡安·米罗（1893–1983）将东方文化中的留白引入西方绘画。“风格派”的发言人杜斯伯格向人们展示了如何在二维绘画中展现三维的空间。密斯·凡·德·罗受到他的

Eastern Culture and Modern Art

Since the extravagant tales of Marco Polo and the merchants' adventures were introduced to Europe in the 17th century, Europe was under the influence of so-called "Chinoiserie."

Chinoiserie was one of the powerful and consistent trends in Europe. It has affected almost all the areas of design, such as décor, furniture, garden follies, tableware and tapestries.

In the 19th century, Europeans opened their eyes to Oriental paintings. It is such a famous story that Van Gogh (1853-1890) was inspired by the colors of Japanese folk paintings. Whereas Van Gogh adopted Japanese style colors, Piet Mondrian (1872-1944), Theo van Doesburg (1883-1931) and Joan Miro (1893-1983) intro-

画作《俄罗斯舞蹈的韵律》（1918年）的启发设计了乡间砖宅（1923–1924年）。此后又受到欧洲抽象艺术家亚历山大·考尔德（1897–1976），在他的雕塑作品中展现出全新的空间价值。

考尔德之前的雕塑作品都是为了表达某个主题而制作的实体，没有一个雕塑展现出空间的价值。然而考尔德的早期作品，就像《美杜莎》（图46），表现出东方文化特有的空间和模糊的边界。

图46
亚历山大·考尔德1930年雕塑作品《美杜莎》
材料：铁丝，尺寸：31cm×44cm×23cm

Fig. 46
Alexander Calder, *Medusa*, 1930, wire, 31cm x 44cm x 23 cm

duced the Oriental value of the void to Western painting. Doesburg, a spokesman of the De Stijl group, showed how two-dimensional paintings were transformed to three-dimensional space. His *Rhythm of a Russian Dance* (1918) inspired Mies van de Rohe to design the *Brick Country House* (1923-24). In later times, influenced by European abstract artists, Alexander Calder (1897-1976) showed a new concept of void in his sculptures.

Before Calder, the purpose of sculpture was to make solid pieces that could express a theme. There was no artwork that showed void space within it. The early works of Calder, however, such as *Medusa* (Fig.46), shows a sense of void and ambiguous boundary, the main characteristics of Eastern culture.

《美杜莎》是利用铁丝展现在三维空间中的作品。观众在欣赏作品时，不能只看到实体物铁丝，只有认识和理解铁丝之间的空间，才能联想到美杜莎。在这件作品中，空间与中国画中的留白一样重要。

考尔德说，他创作“会动的雕塑”是为了让蒙德里安的画动起来。考尔德的“会动的雕塑”是西方美术史上首次把“关系性”当作主题的艺术作品。这些“会动的雕塑”用铁丝把几个物体悬挂在空中，微风吹过便摇摇晃晃，物体之间的距离和角度时刻发生变化（图47）。作品中的这种不断变化的关系远比传统的实体雕塑作品更为重要，这也是考尔德“会动的雕塑”最重要的特点。他的作品从一开始就没有考虑黄金分割律，至少黄金分割律并不重要；相反，东方文化的特征——“虚空”和“关系”才是这些作品中最重要的元素。

The *Medusa* was made of metal wire in three-dimensional space. People cannot perceive the *Medusa* without recognizing void space between wires. The void in this sculpture is as important as the margins in Oriental paintings.

Calder said that he created mobile sculptures with an intention to make Mondrian's paintings move. Calder's mobile sculpture is an art piece which dealt with the theme of "relationships" for the first in Western art history. Several objects made of wire are suspended in the air, and they move whenever the wind blows. The distances and angles among the pieces are constantly changing as time passes (Fig.47). The changing relationships among the pieces are more important than the sculpture made of solid pieces, and this is the most important characteristic of Calder's mobile sculpture. The "Golden ratio" is not considered, or at least is not an important issue, here. Void and relativity, characteristics of Eastern culture, are the most important factors in Calder's works.

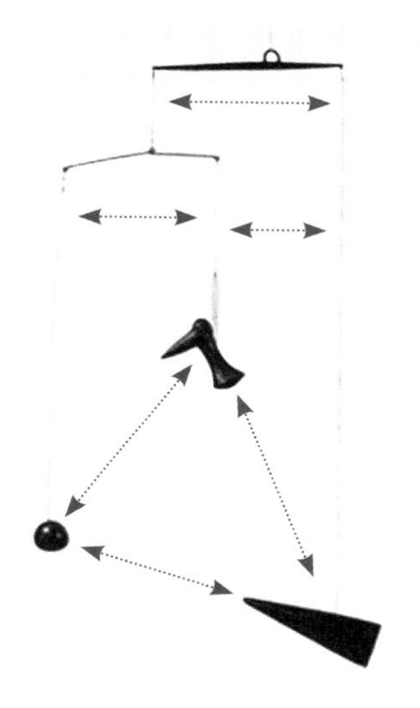

图47
亚历山大·考尔德1933年完成的雕塑——《圆锥》，由乌檀木、金属棒、铁丝组成，尺寸270cm×140cm×61cm。随着观看距离和时间的变化，雕塑的形状会不断改变，形成新的比例和位置关系。

Fig. 47
Alexander Calder, *Cone d'ebene*, 1933, ivoryebony, metal bar, and wire, 270cm x 140cm x 61 cm. The shape changes according to the distance and time, creating a new relationship and proportion.

在保罗·克利（1879–1940）的作品《双向》（1932年，图48）中，没有明确的边界，所有界线都是模糊而重叠的。画中表现出的空间特点与东方建筑的空间特点非常相似。此外，蒙德里安的作品也传递出此类空间感。蒙德里安与杜斯伯格是亲密的朋友，两人一道开创了风格派艺术，他们的作品都呈现出边界模糊的特点。因此，深受杜斯伯格影响的密斯·凡·德·罗的建筑空间具有模糊的边界线就不足为奇了。

弗兰克·劳埃德·赖特与日本

凡·高从日本浮世绘中获得灵感，弗兰克·劳埃德·赖特（1867–1959）也受到日本建筑和绘画的影响。赖特非常喜欢日本的艺术品，他本人就是美国最大的日

The boundaries in Paul Klee's *Two Ways* (1932, Fig.48) are ambiguous and overlapping. The feeling of space in this picture is very similar to that of Eastern architecture. Mondrian's painting style also conveys a similar feeling of space. Mondrian is one of the major figures who started De Stijl with Theo van Doesburg. They were close friends, and their works commonly show ambiguous boundaries. It is no wonder that the architectural space of Mies van der Rohe, who was influenced by Doesburg, has an ambiguous boundary.

Frank Lloyd Wright and Japan

Van Gogh was inspired by Japanese painting, and Frank Lloyd Wright was influenced by Japanese architecture and painting. As the biggest art dealer who sold Japanese paintings in the United States, Wright loved Japanese paintings. The facts that his perspective drawing shows the beauty of void, and the subject and frame are connected to make the boundary between the subject and background

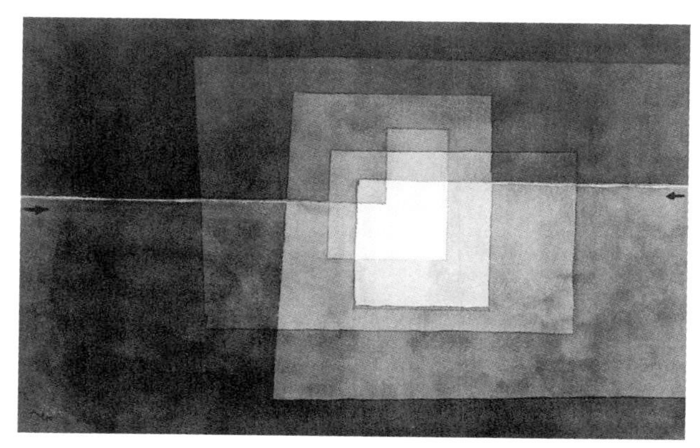

图48
保罗·克利作品：《双向》
（1932年，纸上水彩，44cm×61cm）

Fig. 48
Paul Klee, *Two Ways*, 1932,
watercolor on paper, 44 x 61 cm

图49
桂离宫（1615—1663年）

Fig. 49
Katura Villa, 1615-1663

本艺术品经销商。事实上，他所绘的建筑表现图充分展现出留白之美，画面中的主体与周边有机融合，过渡平缓，使主体与背景融为一体，这些都似乎是受到了东方文化的影响。但赖特的建筑并不具有东方空间的特点，这是因为赖特固守基于墙壁的空间构筑方式。

一旦墙壁成为主要的结构材料，墙体成为划分室内外空间的、泾渭分明的界线。因此，可以说赖特所刻画的空间深深地扎根于西方传统的建筑空间。但密斯和柯布西耶大胆地冲破了西方传统的以墙为导向的空间构筑方式。密斯使用钢柱，柯布西耶使用钢筋混凝土柱作为建筑的主要结构，因此两人所塑造的建筑空间成为流动空间，具备东方建筑的空间特征。

become ambiguous, seem to be influenced by the East. In the architecture of Wright, however, there is no Oriental style void space. It is because he stuck to a wall-based construction principle when he designed space.

Once a wall is used as a main structural material, the wall, as a rigid boundary, clearly divides the inside and outside. The space depicted by Wright is still deeply involved in the traditional Western style space. Mies and Corbusier, on the other hand, daringly broke from the traditional wall-oriented space. Mies used iron columns and Corbusier used concrete columns as a main structural material, and as a result, their architectural space seems to flow like space in Eastern architecture.

图50
蒙德里安作品:《构成》(1916年)
画面的边界模糊。

Fig. 50
Piet Mondrian. *Composition*, 1916
This painting has an ambiguous boundary value.

现代建筑大师的建筑空间分析

The Analysis of the Architectural Space of Two Major Modern Architects

密斯·凡·德·罗的建筑空间演化

众所周知，密斯·凡·德·罗收藏了许多中国书籍，他的书柜上陈列着孔子和老子的著作。密斯的两位挚友——弗兰克·劳埃德·赖特和雨果·贺临（1882–1958）对东方建筑都有着深刻的理解，这使得他有了许多机会接触到东方建筑。而东方哲学研究者卡尔弗里德·格拉夫·杜克海姆（1896–1988）在包豪斯的演讲和会面，决定性地加深了密斯对东方建筑的认识。虽然密斯没有明确表达过自己受到东方建筑的影响，但在1937年接待中国建筑师李承宽的来访时，密斯承认自己的建筑受到中国文化的影响。

密斯的早期作品如克勒拉公寓中仍然可以找出卡尔·弗里德里希·辛克尔（1781–1841）的影响。1910年代的两场展览却完全改变了他，一个是弗兰克·劳

The Evolution of the Architecture of Mies van der Rohe

As a famous collector of Chinese books, Mies van der Rohe had in his study the books written by Confucius and Lao Tzu. His friendship with Frank Lloyd Wright and Hugo Häring, who had a profound knowledge of Eastern architecture, also gave him many chances to encounter Eastern architecture. Decisively, the meeting with Karlfried Graf von Durkheim, who visited the Bauhaus as a speaker, deepened the influence of Eastern architecture. Even though Mies did not directly declare that he was affected by Eastern architecture, he acknowledged the Chinese influence on his architecture when the Chinese architect Chen-Kuen Lee visited him in 1937.

His early works, such as the *Kröller* project, revealed that his designs were still under the influence of Karl Friedrich Schinkel. Two exhibitions in the 1910s, however, completely changed him: They were the exhibitions of Frank Lloyd Wright in Berlin

埃德·赖特在柏林的作品展，另一个是1914年德意志制造联盟在科隆展览会上展示的布鲁诺·陶特（1880–1938）的玻璃展馆。通过这两场展览，密斯从赖特处学会了如何使用钢材，从陶特处学会了如何使用玻璃。在早期的现代主义建筑师中，赖特和陶特正是受日本建筑影响的两位代表。显然，密斯从这两名建筑师处间接地吸收了东方建筑的特质。

赖特试图通过层层墙体实现内外空间的交融。正是因为采用墙体承重的结构，使他的建筑无法演绎出东方建筑特有的那种流动空间。密斯试图在1923年乡间砖宅的设计中尝试一种新的风格，但是仍未能突破赖特的局限。然而，在6年后的1929年，当他设计巴塞罗那博览会德国馆时，他摒弃了承重墙，采用钢柱组成的柱网结构，创造出了连绵的、真正的流动空间。

and the Cologne exhibition in 1914 which introduced Bruno Taut's *Glass Pavilion*. From these exhibitions, he learned how to use steel from Wright and how to use glass from Taut. Wright and Taut are representative architects who were influenced by Japanese architecture in the early period of modern architecture. It is no wonder he indirectly absorbed the characteristics of Eastern architecture.

Wright tried to create an interaction between the interior and exterior space by setting up layers of walls. But since his architecture was still based on wall structure, this structural method held him back from achieving the real fluid void of the East. When Mies tried a new style in his *Brick Country House* in 1923, he could not overcome the limitation of Wright. Six years later, however, he completely created fluid void in *the Barcelona Pavilion* by using grid steel columns instead of structural walls.

1. 乡间砖宅（1923—1924年）

密斯的乡间砖宅深受风格派画家特奥·凡·杜斯伯格的影响。其平面与杜斯伯格的油画《俄罗斯舞蹈的韵律》线条间充盈动态的空间相比，共同点显而易见。密斯的平面图中既没有具方向性的轴线也不具对称性，开放的平面使建筑领域得以扩张，恰如围棋棋盘开放延伸的平面，墙壁自由地延伸，互相交融创造空间。

然而，这两个作品之间还是存在着很大的差异。《俄罗斯舞蹈的韵律》（图51）的空间边界模糊，而密斯在将二维空间转化为三维空间的过程中，仅仅将杜斯伯格画中的线转换为建筑中的墙，却遗失了空间的流动性和边界模糊的特质（图52）。他的建筑看上去好似杜斯伯格的绘画，但实际上他使用的仍是西方传统的墙承重结构体系，他所创造的只是一组内外空间分明的盒子而已。

1. Brick Country House (1923-24)

The *Brick Country House* seems to have been influenced by Theo van Doesburg (1883-1931), a member of De Stijl. His *Rhythm of a Russian Dance* (Fig.51) shows much fluid void space between the lines. The similarity between this painting and *Brick Country House* is easily discernible. The plan shows neither a prominent one-directional axis nor symmetry. Instead, the house expands its territory by adding another piece of wall. Like the pattern of Go, it is an open plan that expands. The walls are spread freely, interacting with each other to create void space.

However, there is a significant difference between two works. Whereas the boundary of the void in *Rhythm of a Russian Dance* (Fig.51) is not clear, Mies failed to convey the essence of fluidity and ambiguity of the void when he translated the two-dimensional void into architectural void space. He simply transformed the lines of the painting into walls of the house (Fig.52). Even though he

图51
特奥·凡·杜斯伯格
《俄罗斯舞蹈的韵律》(布面油画，1918年)

Fig. 51
Theo van Doesburg.
Rhythm of a Russian Dance, 1918, oil on canvas

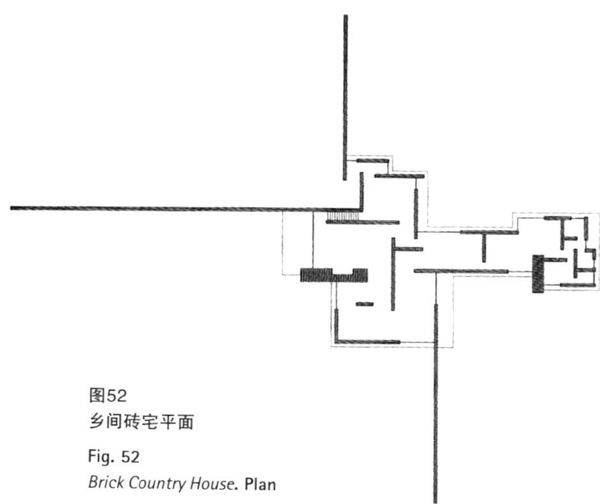

图52
乡间砖宅平面

Fig. 52
Brick Country House. Plan

图53
密斯的乡间砖宅（1923—1924年）

Fig. 53
Mies van der Rohe. *Brick Country House*, 1923-1924

created a house which looks like Doesburg's painting, he still used traditional Western bearing walls as its structural system, creating just an assembly of boxes with clear boundaries between the inside and outside.

2. 巴塞罗那博览会德国馆（1929年）

在巴塞罗那博览会德国馆中，密斯最后采用的是东方建筑常用的结构体系——柱网结构。因此，它的空间特征与东方建筑极其相似，具有自由的流动性。

得益于柱网结构，墙体不再拘泥于轴线对称或是几何形的约束，可以自由地布局。因此，巴塞罗那博览会德国馆深远的挑檐形成“过渡空间”暧昧的边界和模糊的室内外空间，由此营造出流动的空间。

在乡间砖宅中，屋面由墙体支撑，并与墙体相交呈直角，而巴塞罗那博览会德国馆的屋面是悬臂结构，屋檐下的过渡空间形成模糊的边界。

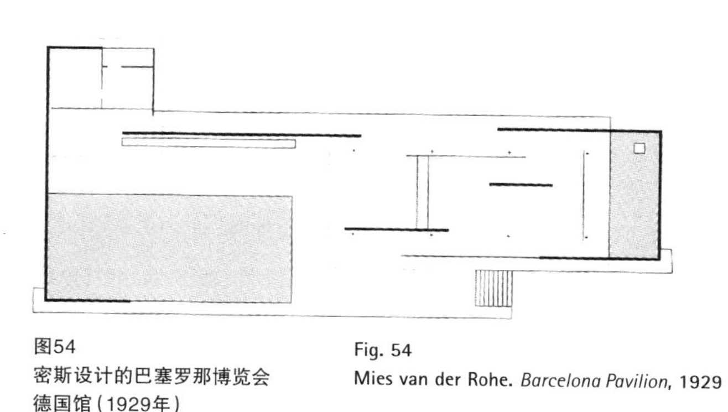

图54
密斯设计的巴塞罗那博览会德国馆（1929年）

Fig. 54
Mies van der Rohe. *Barcelona Pavilion*, 1929

2. Barcelona Pavilion (1929)

From this work, Mies at last adopted a column structure, a main architectural structure in the East. As a result, the space of *Barcelona Pavilion* (1929) looks very similar to that of Eastern architecture, with its free-flowing fluidity.

图55
巴塞罗那德国馆

Fig. 55
Barcelona Pavilion

Thanks to the column system based on grids, the walls are free from strictly bilateral symmetry or geometry. The space of *Barcelona Pavilion*, therefore, has a flexible feeling of space with a vague boundary that makes the inside and outside ambiguous. It is mostly because of the "in-between space" which is made by the roof being extended longer than the building.

图56
京都大德寺（1319—1589年）

Fig. 56
Daitokuji, 1319-1589, Kyoto

The end of the roof of the *Brick Country House* meets the walls at right angles. However, the roof of *Barcelona Pavilion* has a cantilever structure. The space under the eaves becomes the "in-between space" to create an ambiguous boundary (Fig.55).

图57
巴塞罗那德国馆

Fig. 57
Barcelona Pavilion

图58
京都大德寺

Fig. 58
Daitokuji

图59
巴塞罗那德国馆

Fig. 59
Barcelona Pavilion

图60
京都大德寺

Fig. 60
Daitokuji

图61
巴塞罗那德国馆

Fig. 61
Barcelona Pavilion

图62
龙安寺

Fig. 62
Ryoanji

这种过渡空间在密斯后期的代表作品——德国柏林新国家美术馆（1962-1968年）中清晰可见。不同于以墙体为主导的西方建筑，它更像是以屋顶为主导元素的东方建筑。新国家美术馆的设计重点在于屋面的支撑与建造，墙体只用于悬挂艺术品和室内空间的分区。密斯关注于通过屋檐与柱列使建筑成为外部环境景观的一部分，而这正是东方建筑典型的空间特征。

在巴塞罗那博览会德国馆的后院，以不起支撑作用的矮墙隔开庭园，越过矮墙可观看天空和大自然，这与东方式庭园几乎相同。参照图59，图61，图64，可以看出，密斯在巴塞罗那博览会德国馆中使用的柱子结构与东方建筑中柱子组成的网架结构一脉相承。

This "in-between" space is clearly seen in his representative work of the later period *Berlin's New National Gallery (*1962-68). Unlike most wall-oriented Western architecture, this roof-oriented building is more like Eastern architecture. The main architectural task is to build and support the roof, and the walls are just partitions on which to hang artworks. Mies focused on creating a landscape scene framed by eaves and columns, which is a typical spatial feature in Eastern architecture.

In *Barcelona Pavilion*, a garden is built by setting low free-standing walls which have no structural role, and the sky and nature are seen beyond the walls (Fig.59, 61, 64). It is not too much to say that it is the same as Eastern garden. This design was made possible because Mies used a column structure in which walls are free from having a structural role. It is similar to the grid-style column-based architecture in the East.

图63
巴塞罗那德国馆

Fig. 63
Barcelona Pavilion

图64
巴塞罗那德国馆

Fig. 64
Barcelona Pavilion

图65
龙安寺

Fig. 65
Ryoanji

3. 哈勃住宅（1935年）

乡间砖宅展示了基于承重墙的西方建筑的空间感，巴塞罗那博览会德国馆展示了东方建筑的空间感，哈勃住宅则体现了东、西方建筑空间的融合。

哈勃住宅的左右两侧以墙体结构承重，中间部分以柱网承重，洗手间、厨房等服务空间以承重墙分隔，起居室、餐厅等生活空间依靠柱网结构，形成开放的平面布局。

在平面图左下角，密斯利用一道矮墙在起居室前围合出庭园。他还在客厅与庭园之间创造出了一条很窄的门廊，形成空间的过渡，如同东方建筑中曲折的连廊。加上钢柱和全玻璃的墙体，演绎出了与外部空间相同的内部空间。哈勃住宅成功地融合了东西方建筑的空间特征。

3. Hubble House (1935)

If the *Brick Country House* exhibited a wall-based Western style and *Barcelona Pavilion* showed an Eastern style void space, *Hubble House* (Fig.66) is a fusion of Eastern and Western void space.

The left and right parts of the house feature a wall structure and the middle part is based on a column structure. While the serving space, such as the restroom and kitchen, is confined by the bearing walls, the served space, such as living and dining space, reveals an open plan that has structural columns.

On the left lower corner, in front of the living room, Mies made a garden framed by a free-standing wall. He also made a narrow porch area as an intervening space between the living room and the garden. This porch area is similar to a narrow huddle veranda space in Eastern architecture. With the walls made of trans-

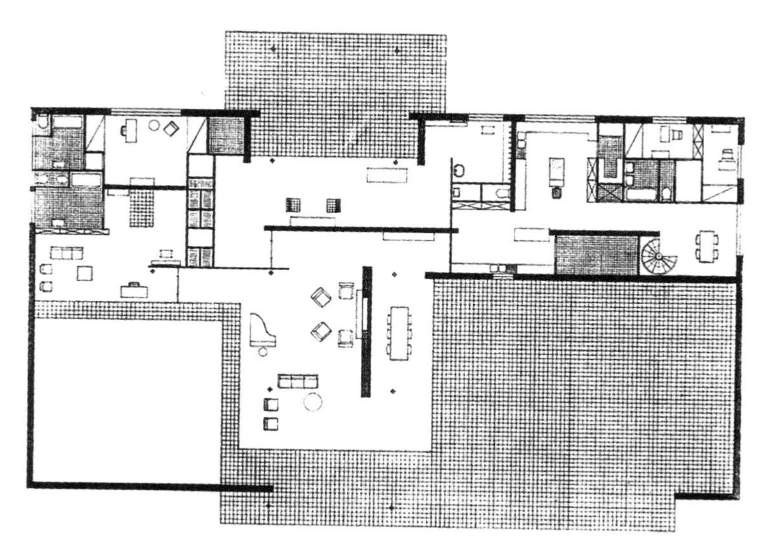

图66
密斯设计的哈勃住宅，1935年

Fig. 66
Mies van der Rohe. *Hubble House*, 1935

parent glass and steel columns, it elicits a feeling of being outside. In terms of void space, the Hubble House is a successful fusion of Western and Eastern architecture.

勒·柯布西耶的建筑空间演化

行文至此，我已经解释了西方建筑空间是基于几何学和数学，以墙体为中心的；相反，东方建筑空间是基于柱网的，以自由平面和流动的空间为特征。这两种迥异的空间特征贯穿勒·柯布西耶的整个职业生涯。

1. 沃卡松别墅（1922年）

这个作品是柯布西耶仍醉心于西方建筑传统的几何学与数学时设计的，别墅的墙体按照精确计算的几何图形布局。

别墅采用的是墙承重的结构体系，墙体分隔出规则的空间，室内外空间也界

The Evolution of the Architecture of Le Corbusier

Thus far, it was explained that the Western architectural space is wall-based, geometric, and mathematical; and Eastern architectural space has a column-based relative structure with grid, free plan and fluid void. These two different types evolved in Le Corbusier's architecture throughout his career.

1. Villa Vaucresson (1922)

This building (Fig.67) was designed when Le Corbusier was still under the influence of typical Western architectural tradition of geometry and mathematics. The walls of the house are located on the mathematically divided geometric lines.

The structure of the house is a bearing wall system; it is a very rigid space divided by walls. The boundary between the interior and exterior is clear, without any fluidity of space or in-between space. His early designs maintained the traditional

限明晰、泾渭分明，没有流动空间，也没有过渡空间。勒·柯布西耶的早期作品坚守着西方建筑的传统价值。不过，他通过塑造极简主义风格的建筑空间将自己从古典主义的建筑装饰中解放出来。

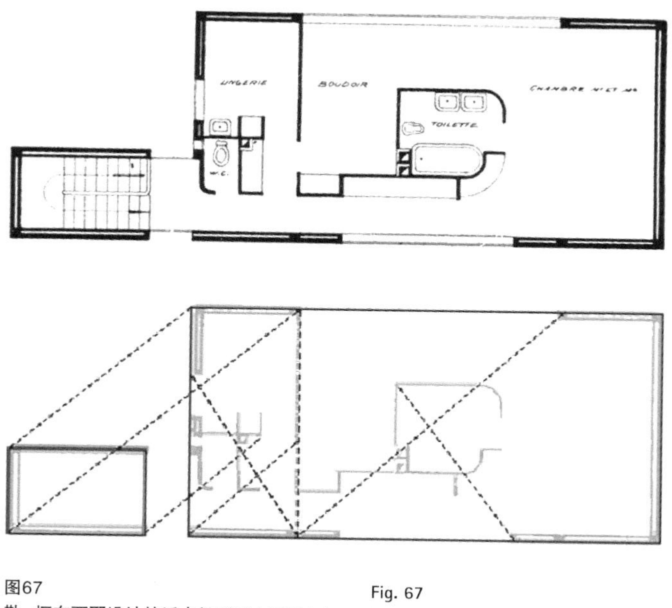

图67
勒·柯布西耶设计的沃卡松别墅（1922年）
四层平面图

Fig. 67
Le Corbusier. *Villa Vaucresson*, 1922.
Third floor plan

values of the West. However, he freed himself from the classic decorative architecture by creating a minimalist style space.

2. 萨伏伊别墅（1929年）

1929年，勒·柯布西耶在《新建筑五点》一书中提到柱网的重要性。同年，他在萨伏伊别墅的设计中采用了柱网结构（图68，图69）。

虽然在萨伏伊别墅中使用了钢筋混凝土柱网的框架结构，但为了追求几何形的平面，柯布西耶改变了几处柱子的位置。这是几何学的价值凌驾于柱网结构之上的证据。萨伏伊别墅的柱网体系是5×5共25个柱子，其中10个柱子的位置或是变更、或是分置、或是以承重墙代之。可以想象，在进行平面设计时，柯布西耶先设计了柱网，再在矩形柱网区域布置由圆形与矩形房间组成的平面，矩形房间的墙体取代了几根柱子。即使这样，由于采用了柱网结构，在各个功能空间之间形成许多过渡空间。

2. Villa Savoye (1929)

In his book *Five Points of the New Architecture* (1926), Le Corbusier mentioned the importance of the grid column system. In 1929, he used a grid column structure in his design of *Villa Savoye* (Fig. 68, 69).

He introduced a concrete column grid, which was modified to pursue a geometric plan. It shows the fact that the geometric value is given priority over the grid system, yet out of 25 columns on the 5 x 5 grid column system, 10 were relocated, split, or replaced by bearing walls. It is assumed that he first laid out the column grid and overlaid the plan of geometric circles and rectangular boxes within the realm of the rectangular grid system when he designed the plan. The walls of the box-like rooms were used as substitutes for some columns. However, the grid column structure created the so-called "in-between space" with multiple features in some parts.

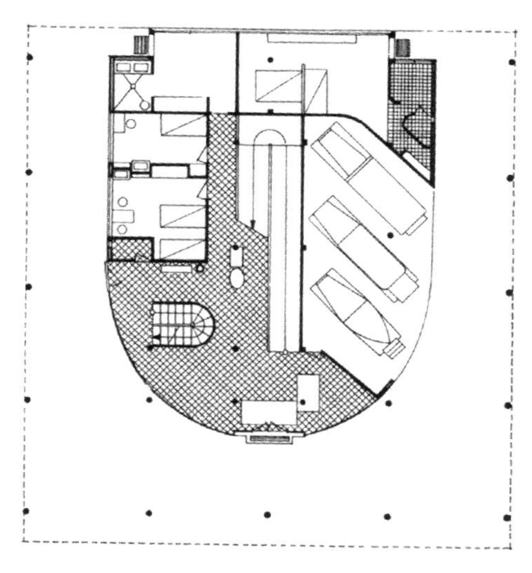

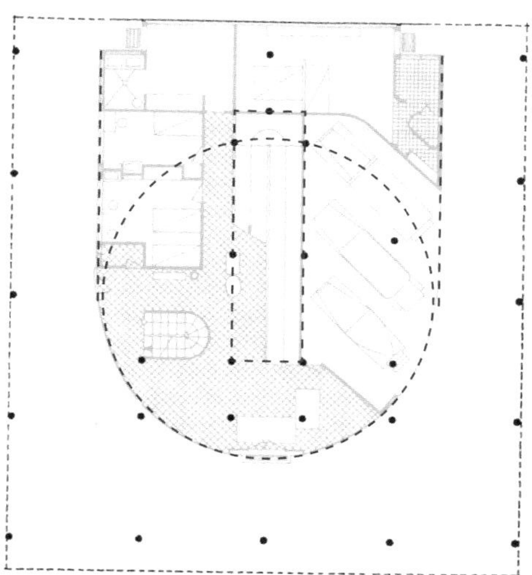

图68
勒·柯布西耶设计的
萨伏伊别墅（1929年）
底层平面图

Fig. 68
Le Corbusier. *Villa Savoye*, 1929.
Ground floor plan

图69
萨伏伊别墅

Fig. 69
Villa Savoye

图69中，室外景观与室内空间之间的过渡空间——一层架空的柱廊以混凝土厚板为顶，这块厚板也是二层的楼板。连续的柱子界定出了空间。这个空间很像雅典帕提侬神庙的门廊空间，或者如柯布西耶自己说的，像汽艇的甲板。

图70
Yoshimura-tei

Fig. 70
Yoshimura-tei

As seen in Figure 69, the space between landscape and the interior is covered with a slab, which constitutes the second floor; and the serial columns claim this territory. This space is similar to the porch space in the *Parthenon*, or, as Le Corbusier said, a steamboat deck.

因为有连续的墙面，无论帕提农神庙还是汽艇，室内外的空间界线始终是分明的。而萨伏伊别墅的过渡空间——檐廊（图71），更像是东方建筑中木柱构成的檐下空间（图72），它的室内外空间之间的界线是模糊的。似乎从那之后，柯布西耶开始设计东方风格的建筑空间。但他并不拘泥于柱网的阵列布局，仍然会用承重墙来代替一些柱子，曲面墙体采用的也是几何弧形而不是自由曲线。

图73是在日本海滩上建造的严岛神社，为了避免涨潮时被海水浸没，用柱子将地板抬高。不只这个神社如此，为防止受潮，大部分东方建筑的地板都是抬高的。也许是因为东方农耕社会以水稻种植维持生计，而水稻的种植需要丰富的雨水，人们不得不选择住在降雨充沛的地方。因此，让建筑物脱离潮湿的大地理所当然地成为东方建筑必须解决的问题。

However, whereas the boundary between inside and outside is clear in the *Parthenon* and a steamboat because of continuous walls, space of the *Villa Savoye* is rather similar to the eaves which were made using wooden columns in Eastern architecture (Fig.72). This space has an ambiguous boundary between the interior and exterior. It seems that Le Corbusier started to design Eastern style space from this time. However, even though he introduced a grid column, bearing wall structure still was given priority in certain locations; and the curved walls were geometrically shaped instead of being freely curved.

Figure 73 is a *Miyajima Shrine* building built on the foreshore. Its columns raised the floor of the building to prevent it from sinking underwater with the rising tide. Not only this shrine, but almost all the traditional Eastern buildings are raised above the ground. It is because the ground condition of the site in the East was damp. Since the East was composed of agrarian societies where the main liveli-

图71
萨伏伊别墅

Fig. 71
Villa Savoye

图72
Yoshimura-tei

Figure. 72
Yoshimura-tei

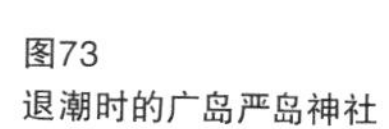

图73
退潮时的广岛严岛神社

Fig. 73
Miyajima near Hiroshima Building of the *Itsukushima Shrine* at low tide.

这个问题的解决之道就是利用柱网结构将建筑抬高，从而与潮湿的地面脱离。柯布西耶正是采用了这一方法，并且是按此解释的（图75）。密斯以同样的方式建造了范斯沃斯住宅（1945-1950年），只不过他们所用的材料不同。柯布西耶用的是钢筋混凝土柱，而密斯则用的是钢柱。

图74
奈良法隆寺慈光院，
移门全部推开时的景象

Fig. 74
Jiko-In near Horyuji, Nara.
Picture taken when all the sliding doors are opened.

hood was rice farming demanding lots of rain, and the people thus had to live in regions with a great deal of precipitation. Therefore, separating the building from the wet ground must be an important architectural challenge in the East.

The solution was to use a column structure to raise the buildings. Le Corbusier

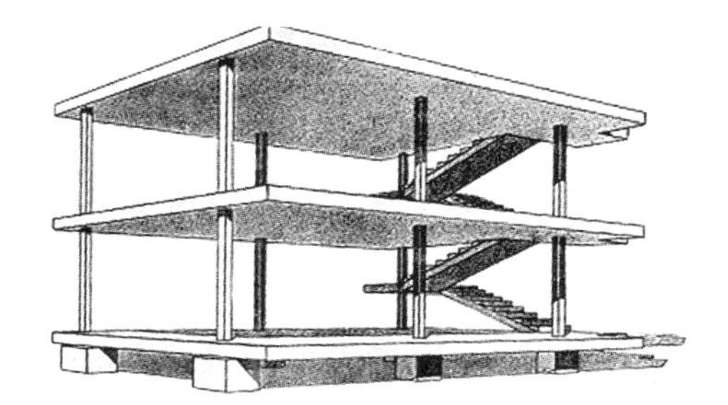

图75
勒·柯布西耶设计的多米诺住宅
（骨牌屋，1914年）

Fig. 75
Le Corbusier. *Domino House*, 1914

图76
密斯·凡·德·罗设计的范斯沃斯住宅（1945—1950年）

Fig. 76
Mies van der Rohe. *Farnsworth House*, 1945-50

adopted this method and explained that the merit of a raised ground floor was to free the house from unpleasant ground moisture. Mies also used the same method in *Farnsworth House* (1945-50). They used different material, though: Le Corbusier used concrete and Mies used steel instead of wood.

3. 米尔住宅（1954年）

柯布西耶从这一作品开始打破了欧几里得几何学的窠臼（图77）。他在矩形柱网平面上建造自由曲面墙，由此在曲面墙体与柱子间形成界线模糊的过渡空间。柱子时而位于曲面空间内，时而位于曲面空间之外，使得空间更具流动性，也使得内外空间得以交融。

柯布西耶在悬挑的混凝土厚板下构造过渡空间，而不是使用檐廊。从平面图上可以看到，平面基于一个5×4的框架结构，但是20根立柱中有12个被承重墙取代，比萨伏伊别墅中的柱网变动还要多。柯布西耶在这一作品中，打破了欧几里得几何学的制约，但是他的自由仅仅是在矩形盒子内的自由，而没能冲出矩形柱网这个“盒子”。

3. Mill Owner's House (1954)

From this work (Fig.77), Le Corbusier began to break away from Euclidean geometry. He set up walls made of free curves within a rectangular column grid. Space between the curved wall and the columns became ambiguous in-between space. The columns are located inside and outside the curved room, creating a fluid space that joins the exterior and the interior together.

He created in-between open space under the slab, instead of using the eaves. The plan is based on a 5 x 4 column grid, and 12 out of 20 columns are replaced by bearing walls. Compared to the *Villa Savoie*, the grid system is less preserved in the *Mill Owner's House*. Even though he broke from the Euclidean geometry in this work, his design did not cross over the boundary of the rectangular box, permitting the free style only within the boxed frame.

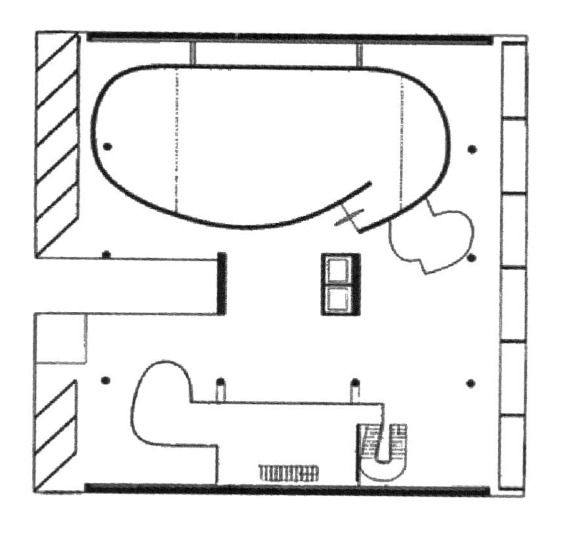

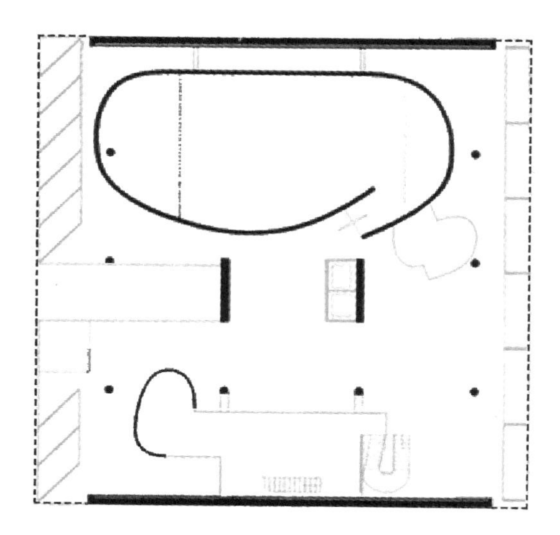

图77
勒·柯布西耶设计的米尔住宅（1954年）

Fig. 77
Le Corbusier. *Mill Owner's House*, 1954

4. 卡彭特中心（1961年）

位于美国马萨诸塞州坎布里奇市哈佛大学校园内的卡彭特中心，是勒·柯布西耶的建筑中最接近东方建筑价值观的作品。

首先，柯布西耶完整地保留了这一建筑所有的柱网框架结构，没有承重墙，所有的墙体都从结构的束缚中解放出来。设计的过程完全不受几何学的制约，自由的曲面墙体可以肆意地布局。柯布西耶之前的作品中，所有的曲面墙体仅仅是局限在矩形箱体内。即便是设计一个流动的弧形空间，他也总是在矩形柱网内进行设计的。卡彭特中心的平面完全脱离了这种局限，彻彻底底地由自由的曲线墙组成。

卡彭特中心的设计基本上侧重于人在时－空四维空间中运动的体验，这与传

4. Carpenter Center (1961)

Located on the campus of Harvard University, Cambridge, Massachusetts, the *Carpenter Center* reveals Le Corbusier's most evolved architectural style, which displays the design closest to Eastern architecture.

First of all, he retained the grid column structure throughout the entire building. Without bearing walls, the walls are totally free from the structure. There is no geometric restriction in this design; and the walls made of free curves are exposed to the outside. In his previous projects, all the curved walls were located within a rectangular box. Even when he designed a curved space with free fluidity, it was always under the control of rectangular geometry. Completely breaking from the limitation, however, the *Carpenter Center* displays a plan totally composed of free curves.

统的西方建筑空间从几何与数学以及旁观者的视角为出发点进行的设计大不相同。它与之前讲到的挣脱了几何学束缚的风景式园林设计极为相似。传统的西方建筑都有一个与入口通道轴线垂直的主立面，当人们走进主入口时可以看到这个二维的立面，它的设计遵循“黄金分割律”。

在柯布西耶初期作品中也可以看到与之相近的设计思维。但是在他的后期作品——卡彭特中心中，他放弃了通过与建筑物正立面垂直的通

The design of the *Carpenter Center* basically focuses on a moving person's perception. It is very different from traditional Western buildings which were designed based on geometry and mathematics from third person's perspective. It is similar to the change that appeared in the picturesque garden design that broke from geometry. Traditionally, Western buildings have their main axial passage perpendicular from the main façades so that a person who enters the passage can see a two-dimensional elevation, which was designed according to the "Golden ratio."

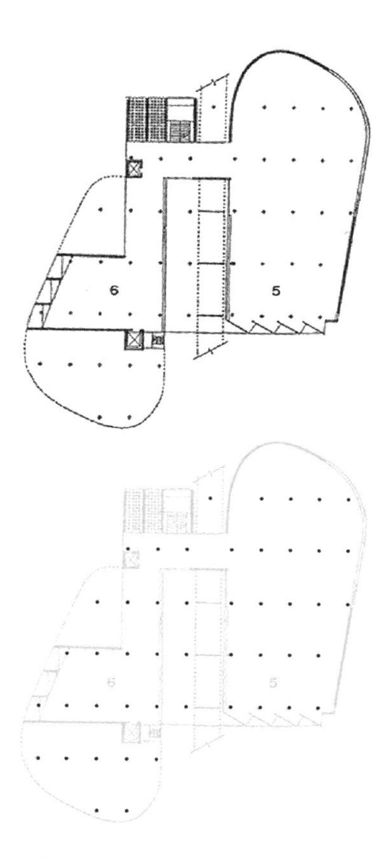

图78
柯布西耶设计的卡彭特中心（1961年）
三层平面图

Fig. 78
Le Corbusier. *Carpenter Center*, 1961
Second floor plan

道进入建筑的方式，代之以与人行道平行的一个斜坡，以弯曲的路径靠近建筑，直到临近入口处才与正立面垂直相交（图79）。由于卡彭特中心没有与正立面垂直的入口通道，所以人们无法瞬间获取对建筑的客观、全面的认知。人们只能得到一些建筑立面侧面的片段，并通过重组这些视觉片段和空间图像组成一个他们心目中建筑的完整形象。这种设计方法与17世纪英国风景式园林的设计方式相似，感性的园林设计方法终于在300年后由勒·柯布西耶运用到建筑设计中。

此外，卡彭特中心的设计强调了空间的重要性。柯布西耶以独特的方式用虚空的空间处理了建筑的水平中心与垂直中心（图80）。在建筑物中，主入口斜坡道贯穿整栋建筑，与建筑对面的人行道相连，建筑的中心得以形成穿堂风。从广义的角度说，卡彭特中心以空间为中心而不是以几何学意义上均衡的正立面为中

A similar design approach was seen in Le Corbusier's early works, too. In the *Carpenter Center*, one of his later works, however, he refused to use the perpendicular approach to the main façade. Instead, he used a ramp passage that starts parallel to the street, curves in approaching the building, and meets the building at a perpendicular angle before the entry (Fig.79). Since there is no main perpendicular axial passage to the main entrance from a distance, there is no chance to perceive the building objectively and generally. People who get only the fragmental perspective views of the building elevation compose a total image of the building in their minds by reorganizing the fragmental and spatial images. This kind of design method is similar to the English picturesque garden design which appeared in the 17th century. This perceptual trend of garden design finally appeared in architecture by Le Corbusier after 300 years.

Also, the design of the *Carpenter Center* emphasizes the importance of the void.

心。为了创造流动的过渡空间，柯布西耶在卡彭特中心采用了柱网结构体系，加之建筑中虚空空间的利用，以及基于第一人称体验的设计，追求设计的相对意义，这一切都与东方建筑的特征是一致的。

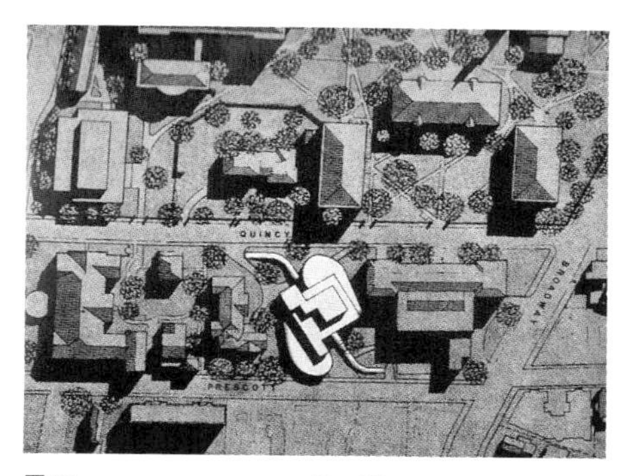

图79
卡彭特中心总平面图

Fig. 79
Carpenter Center. Site plan

图80
卡彭特中心正立面

Fig. 80
Carpenter Center. Façade

When Le Corbusier adopted the void in his architecture, he left the vertical and horizontal center of the *Carpenter Center* as void in a unique way (Fig.80). That is, the main ramp penetrates the building to reach the opposite sidewalk, making the center of the building a ventilated void. Broadly speaking, the *Carpenter Center* created void space instead of a geometrically balanced main façade in the center of the building. Le Corbusier used a grid column as a structural system for the fluidity of in-between space, made full use of void space at the center of the building, and introduced a design based on a first-person perspective to create a relative value. All these characteristics are identical to those of Eastern architecture.

当代建筑大师的建筑空间分析

The Analysis of the Architectural Space of Two Major Contemporary Architects

路易·康：静谧的空间

尽管密斯与柯布西耶将东、西方建筑“杂合”后创造出了现代建筑，其他现代主义建筑师们从这两位大师处所学到的似乎只是国际式风格。现代主义建筑的后半时期完全被国际式风格所主导，那是“形式追随功能”的时代，没有功能的空间一律遭到摒弃。然而，路易·康打破了国际式风格的陈规，在现代建筑中重新引入虚空空间。在康设计的大部分建筑平面中，居于核心的都是虚空空间，而且这些虚空空间都基于西方传统的建筑空间——具有几何形的空间。

图81至图84说明了康设计的虚空空间与西方传统建筑空间的相似之处。几何形态相似的拱券构成反射光，创建了类似的空间。身为犹太人，路易·康在西方传统中融入了犹太传统。

Louis Kahn: Silent Void Space

Even though Mies and Le Corbusier introduced hybrid architecture of East and West, what other modern architects learned from them was only "International Style." The second half period of the modern architecture was dominated by International Style and it was the era when "Form follows function." Void space without specific function in architecture was eliminated. Louis I. Kahn, however, broke from the International Style and reintroduced void space into modern architecture. Most of his building plans were made with void space in the center, and his void is based on the traditional Western geometric space.

Figure 81-84 illustrates the similarity between Kahn's void space and traditional Western architectural space. The reflected light produced by the use of geometric arch structure created similar space.

图81
路易·康设计的菲利普斯·埃克塞特学院图书馆（新罕布什尔州，1965—1972年）

Fig. 81
Louis Kahn. *Phillips Exter Academy Library*, 1965-1972, Exter, New Hampshire

图82
哥特教堂

Fig. 82
Gothic Cathedral

图83
路易·康设计的耶鲁大学英国艺术中心，康涅狄格州纽黑文市（1969—1975年）

Fig. 83
Louis Kahn. *Yale Center for British Art*, 1969-1975, New Haven, Connecticut

图84
哥特教堂

Fig. 84
Gothic Cathedral

犹太人相信某些特殊的几何图案具有一种神秘的力量，所罗门就画了许多这样的图案，大体上是圆内接三角形或矩形，以此得出欧几里得几何的重叠。康设计的许多建筑平面图都具有类似的图案，最典型的案例是“耶鲁大学美术馆”（1951-1953年，图86）的楼梯间。他在圆形的楼梯间内，将楼梯设计为内接的三角形，楼梯间顶部反射自然光线的反射板也设计成三角形。除此之外，康在1953年设计的费城市民中心也有类似的平面（图88）。

康在追随传统的同时，也构筑了自己独特的建筑世界。他既使用西方传统的几何形设计手法，也懂得运用东方枯山水中营造的虚空空间。在康的代表作之一——索克生物研究所（1959-1965年，图89，图90）中，建筑的中心广场最初被设计成了一个郁郁葱葱的庭园。1965年12月到墨西哥城旅行时，康参观了墨

As a Jew, Louis Kahn also integrated Western tradition with that of the Jews. The Jews believed a certain combination of geometry holds a special power, and Solomon designed a lot of such pentacles, most of which have a triangle or a rectangle inside a circle, showing overlapped Euclid geometry. Kahn used Solomon' s symbols in many of his architectural plans, and the most famous example is the staircase of the *Yale University Art Gallery* (1951-53, Fig.86). He made a circle-shaped staircase, put triangle-shaped stairs inside, and installed a reflector that reflects the natural light on the ceiling. A similar plan is seen in his *Philadelphia Civic Center* (Fig.88) designed in 1953.

While following tradition, Kahn built a unique architectural world of his own; he used not only Western geometry but also the Zen type of Eastern void. When he first designed the Salk Institute (1959-65, Fig.86), one of his representative works, the central plaza was originally designed as a verdant garden (Fig.90).

图85
所罗门之匙
(圆内切倒三角形与圆内切五角星)

Fig. 85
The seventh and last pentacle of Saturn from the book of *Key of Solomon*

图86
路易·康设计的耶鲁大学美术馆
(康涅狄格州纽黑文市, 1951—1953年)

Fig. 86
Louis Kahn. *Yale University Art Gallery*, 1951-1953, New Haven, Connecticut

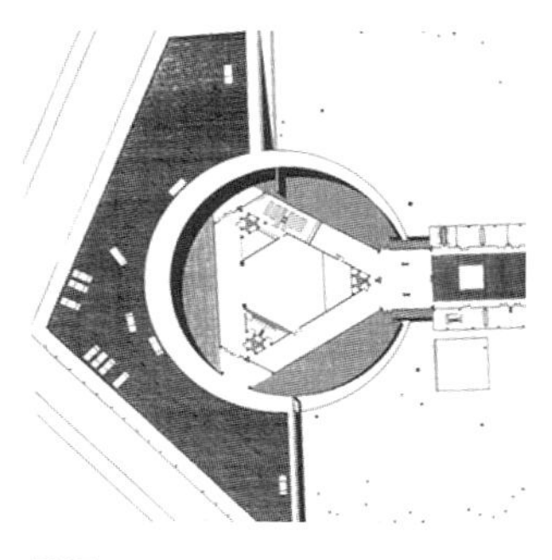

图87
路易·康设计的犹太教堂与学校平面图
(宾夕法尼亚州费城, 1954年)

Fig. 87
Louis Kahn. *Adath Jeshurun Synagogue and School Building*, 1954, Philadelphia, Pennsylvania. Plan

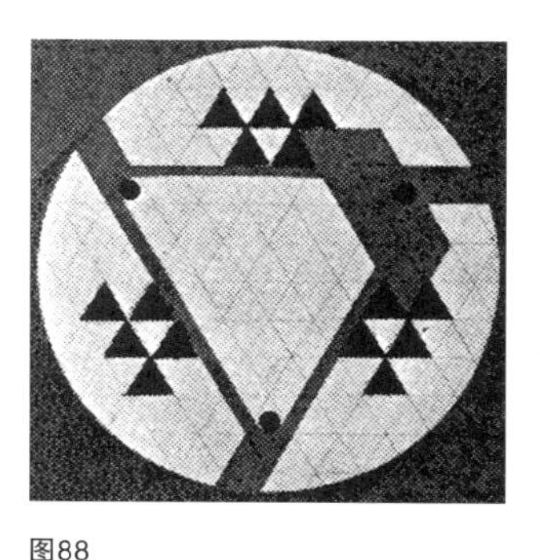

图88
路易·康设计的费城市民中心酒店及百货商店平面图(宾夕法尼亚州, 1953年)

Fig. 88
Louis Kahn. *Philadelphia Civic Center*, 1953, Philadelphia, Pennsylvania. Plan diagram of hotel and department store

西哥建筑师路易斯·巴拉干（1902–1988）风格简约质朴的庭院后，邀请巴拉干到加州，给建造中的索克大学研究所的庭院提些意见。巴拉干如邀来到工地，建议康以铺石的广场代替树木或草坪的庭院，因为“如果你把它做成一个广场，你会获得一个新的立面——一个朝向天空的立面。”

康接受了巴拉干的忠告，将种满花草树木的中庭腾空，改为一个仅以石铺地的空旷的中庭，这就是著名的索克生物研究所的广场（图89）。康从巴拉干处学会了如何让建筑接受自然，就像汉弗莱·雷普顿在阿丁汉大道（图44，图45）中迁移树木得到空白的空间一样，康也在索克生物研究所设计了一个静谧的、具有枯山水意境的广场 。

In December 1965, he visited the austere garden designed by Luis Barragán, a Mexican architect, during the trip to Mexico City, and he invited Barragán to visit the construction site of the *Salk Institute* and to give him some advice about the courtyard. Barragán came to see the site and advised Kahn make a plaza made of stone, instead of planting trees or grasses. He said if Kahn made a plaza, he would gain the sky as a façade.

Kahn accepted the advice and designed the famous plaza of the *Salk Institute*. Barragán's advice changed the Kahn's verdant garden design into a void stone plaza. From Barragán, Louis Kahn learned the way how architecture accepted nature. Kahn could create the silence of a Zen garden by emptying the garden, instead of filling it with trees, just as Humphry Repton removed trees in his *The Drive at Attingham* (Fig.44, 45).

图89
路易·康设计的索克生物研究所
（加利福尼亚州，拉霍亚市，1959—1965年）

Fig. 89
Louis Kahn. *Salk Institute for Biological Studies*,
1959-1965, La Jolla, California

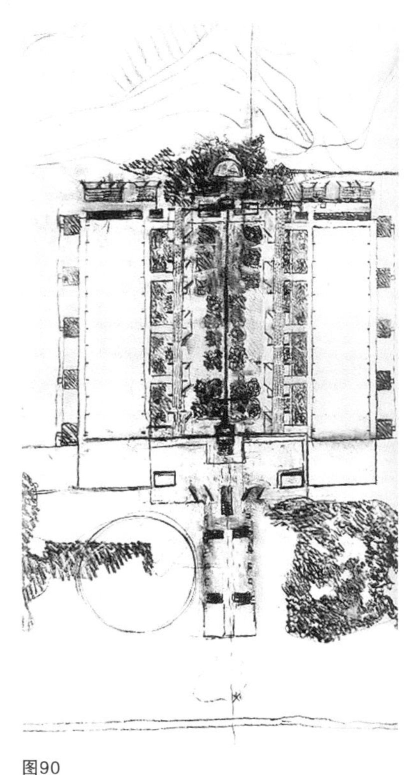

图90
路易·康设计的索克生物研究所
最初总平面草图（1962年）

Fig. 90
Louis Kahn. *Salk Institute for Biological Studies*,
1962, original site plan sketch

道可道，非常道。名可名，非常名。无名天地之始。有名万物之母。[4]

老子《道德经》第一章

据《静谧与光明》一书作者约翰·罗贝尔的描述，路易·康所说的静谧与老子的"无名"相同，康说的光明即老子的"有名"。概括地说，康作为建筑师成功地将东方静谧的虚空空间加入西方几何性的结构体系中。

Lao Tzu said, "*The truth that can be told is not the eternal truth. The name that can be named is not the eternal name. The nameless is the beginning of the heaven and the earth, and the named is the mother of the ten thousand things.*" [4]

In the book *Between Silence and Light* written by John Lobell, what Kahn called "silence" is "the nameless" in Lao Tzu's poem; what Kahn called "light" is "the named." In a nutshell, Kahn is an architect who successfully fixed the concept of void space of the East into the geometric framework of the West.

[4] Lao Tzu. *Tao Te Ching*. Ch.1

安藤忠雄：东西方建筑的融合

安藤忠雄的建筑空间兼具东方的相对关系和西方的几何性。他的建筑创造了一系列步移景异的空间效果。这与英国风景式园林和东方传统园林的设计原则是相似的。安藤说过："我想做出这样的建筑空间，身处其中的人们可以静静地感受，但绝不声张自己的感动。"[5] 也许正是由于这一原因，在他的建筑中，有路易·康和日本传统建筑中那种静谧的虚空空间。安藤将康和柯布西耶作为自己的精神导师、激励自己自学建筑的故事众人皆知。他曾想拜柯布西耶为师，可当他抵达欧洲时，柯布西耶已经去世了。安藤回到日本后，通过研习柯布西耶的建筑图纸自学建筑设计。出于对柯布西耶的深深怀念，安藤给自己养的宠物小狗起名为"柯布西耶"。从遗传谱系的角度看，安藤忠雄所学的正是路易·康和柯布西

Tadao Ando: Hybrid Architecture of the East and the West

Tadao Ando's architectural space is based on Eastern relativity and Western geometry. His buildings are the tools that create a perspective view that changes according to the viewer's location and a changing relationship among peripheral elements as a result. It is quite similar to the principles of picturesque garden design and the Eastern stroll garden. Ando once said, "*I want to make space in which people are so quietly moved that they don't make a fuss about it.*"[5] Accordingly, his buildings have "silent" void space as in Kahn's architecture and traditional Japanese architecture. It is such a famous story that Ando, a self-educated architect, benchmarked Corbusier and Khan as his architectural mentors. It was after Corbusier died when Ando went to Europe to learn from his architectural works. After he went back to Japan, Ando studied architecture by himself by tracing all

[5] Francesco Dal Co, *Tadao Ando Complete Works*. (London: Phaidon, 1994) p. 500.

耶的东西方文化杂合后的建筑设计。

要理解安藤的建筑空间，应该先了解日本的传统建筑，特别是16世纪茶道大师千利休(1521-1591)，他对安藤的建筑风格和建筑空间有很大的影响。作为最伟大的茶道大师之一，千利休在传统建筑的基础上略加改动后设计出的茶室，为茶道注入了鲜活的生命力。

在千利休的建筑中，入口通道是非常重要的元素，来访者沿着这条入口通道得以体验一连串如画的景色。这种线形的入口通道在安藤的许多建筑中也可以见到，与之不同的是安藤建筑中的入口通道，或水平，或垂直，更为曲折。人们走过曲折的通道，可以获得多种视角的建筑景象，这种从多个视点收集不同的视觉片断，从而体验到三维空间的方法，与土木工程师测量地形时所用的"三角法"

the architectural drawings of Corbusier. He even named his dog "Corbusier." Considering the hereditary genealogy of his architecture, it is natural that he learned hybrid architecture of the East and West from Corbusier and Kahn.

In order to understand Ando's architectural space, we need to start with Japanese traditional architecture, especially that of 16th century tea ceremony master Senno Rikyu (1521-91) who had a great deal of influence on Ando's architectural style and space. As one of the greatest tea ceremony masters, Rikyu brought fresh vitality into the tea ceremony by designing small pavilions that were slightly different from previous traditional architecture.

The approaching path before the main space is a very important key to Rikyu's architecture. Walking along the path, visitors get to have a sequential picturesque experience. This linear approach path is shown in many of Ando's architectural

的原理相同。

工程师在测量地形时，通过水平尺刻度得知高低之差；在安藤的建筑中，人们通过台阶级数以及间或的不同高差的平台，感受自己在垂直方向上的运动。这种空间手法在安藤的两个建筑作品中有所体现：水之教堂与风之教堂。

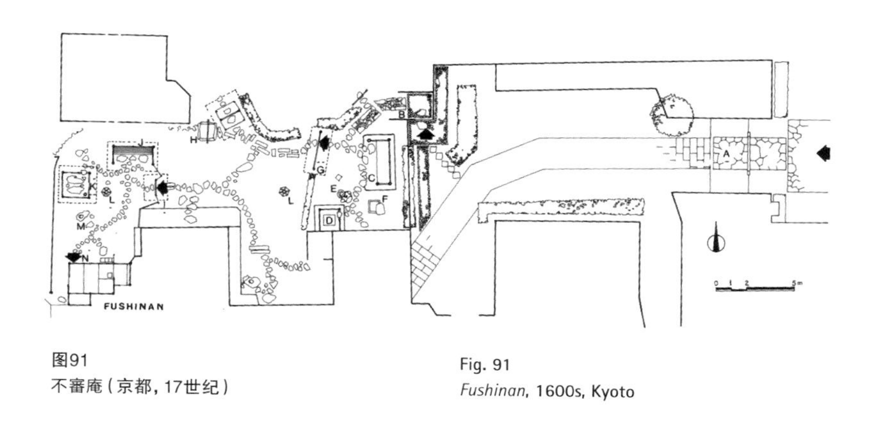

图91
不審庵（京都，17世纪）

Fig. 91
Fushinan, 1600s, Kyoto

designs. Ando's buildings are perceived by the sequence of picturesque perspectives from the linear path, which twists and turns horizontally and vertically. In this way, the person on this path gets to experience different perspective plans from various viewpoints, just as civil engineers do land surveys using a triangulation method.

While civil engineers read a leveler to measure the difference of elevation, people feel their vertical movement by counting the number of steps or occasional stair risers in Ando's buildings. This spatial technique can be seen in two of Ando's works: One is the *Chapel on the Water* and the other is the *Chapel on Mt. Rokko*.

水之教堂（1985—1988年）

测绘工程师在测量地形时，需要在某一地点收集完必要的数据后，再移到另一地点收集新的数据。通过观察者在不同位置上得到的数据，结合其在二维平面上移动的距离和移动的角度，可以计算出三维的地形尺寸。这种方法叫做三角测量。

同样，当人们从水之教堂的后面走近它时，对教堂及其周边的环境空间会逐渐产生印象。人们首先会看到一道混凝土围墙和主体建筑，一座由四个混凝土十字架以及以钢制竖梃构成的玻璃立方体。沿着入口通道走到混凝土墙前，视线被墙遮挡，眼前所见只有一道墙。沿着入口通道继续向前走，左转，视野左侧出现了自然景观，而右侧仍是混凝土墙。由自然和混凝土构成的场景伴随着入口

1. Chapel on the Water (1985-88)

A land surveyor gets the necessary data at a point and moves to another point to collect the data of that point. It is triangulation method to figure out three-dimensional topography by calculating the data obtained at different locations and the information on how far the surveyor moved in what angle on the two-dimensional drawing.

Similarly, visitors recognize the building and the nearby views as they approach the *Chapel on the Water* from the rear. First, they see a freestanding concrete wall and a glass tower of the main building, which is composed of four concrete crosses and four metal mullions in a cubic geometry. As they follow the passage, they get to stand in front of the concrete wall. What they see is just a concrete wall. And when they continue along the passage, bending to the left, they see the landscape on the left and the concrete wall on the right. The composition of

图92
安藤忠雄设计的水之教堂
（北海道勇払郡，1985—1988年）

Fig. 92
Tadao Ando. *Chapel on the Water*, 1985-88, Tomamu, Hokkaido

通道一直延伸到混凝土墙面的尽头。

走在长长的通道上，人们可以感受到建筑基地周围环境与建筑之间的关系。进入大门，访客能看到一个45m×90m的人工水池，池中竖立着金属的十字架，还有教堂。水、十字架和教堂三个空间元素之间的关系一目了然。

从左到右，依次可以看到水、十字架和教堂。顺着小路朝混凝土围墙的对面走，只需几分钟就能够看到这一空间迷宫的第二部分。在墙的端头，向左转90°，抬头望到的是此前在空间之旅开始时就看到的玻璃方盒子。

然后，再左转走进一个狭窄的楼梯间。身处混凝土墙面包围的楼梯间，人与周围的环境完全隔离。此刻正是进入下一个如画风景的瞬间之前的短暂静谧。在这里，只有混凝土空间及自己。随后，走上楼梯，就来到整个建筑的最高

landscape and the concrete wall continues until they reach the end of the wall.

Walking on the long path, they begin to understand the relationship between the surrounding site context and the building location. After passing the gate, they see a 45m x 90m man-made pond, a metal cross in the middle of the pond, and a main chapel building. They capture the relationship among these three elements of the chapel from a relatively objective point of view.

From left to right, they see a pond, a cross, and a chapel. Walking along the path on the opposite side of the freestanding concrete wall that they just walked along a few minutes ago, they figure out the second piece of this space puzzle. At the end of the wall, they turn 90 degrees to the left and see the glass cube that they saw at the beginning of this journey.

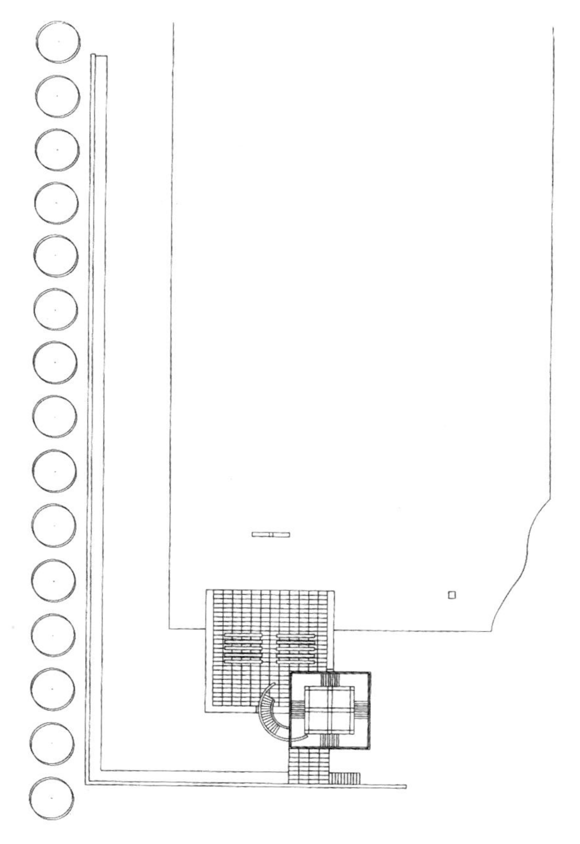

图93
水之教堂总平面图

Fig. 93
Chapel on the Water. Site plan

点——玻璃盒子。

沿着螺旋式的楼梯上行，人们不知不觉中旋转了360°，经历的所有景象被全景式地重温。犹如一个指南针，几何性的玻璃立方体为人们提供了理解周边自然景观与复杂建筑的三维网格参照，脑海中支离破碎的生动画面得以重新整合，构成了一个完整的图像。

人们有机会再次一览混凝土围墙、水池、金属十字架及周围景观，这也是第三个可以一窥全貌的观看点，前两个分别是入口通道前的出发点，以及转过混凝土围墙的瞬间。人们从沿着混凝土围墙走，到之后反转180°再次沿着墙的背面走，因为距离相同，方向相反，所以可以清晰地感受到起点的距离及方向。

所有设计复杂的历程都是为了让人们得到丰富的空间体验。在水之教堂的

Then, they turn to the left and enter a narrow staircase. On these stairs, they are totally separated from the surrounding context. It is a blank moment before the next picturesque moment. At that moment, there are only concrete-framed space and themselves. After that experience, they ascend to the highest point of this building, the glass cube.

On this staircase, they rotate 360 degrees and take a moment to look back at all the sequence they went through from an omniscient point of view. Like a compass, this geometric glass cube is a three-dimensional grid reference of this landscape and building complex. They reorganize these fragmented picturesque images in their minds to get the whole picture.

Once again, they look at the freestanding concrete wall, the pond, the metal cross, and the surrounding views. This is the third surveying point. The first was

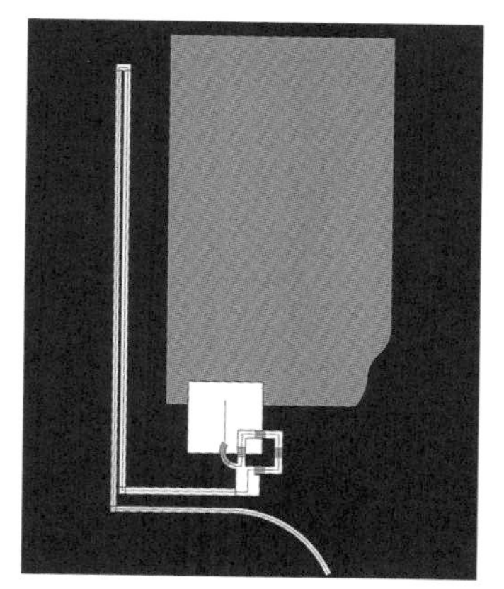

图94
入口通道俯视图

Fig. 94
Top view of
the approach path

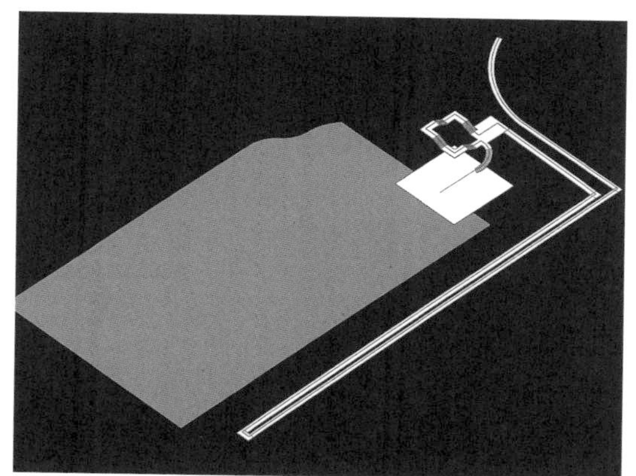

图95
轴测图

Fig. 95
Axonometric view

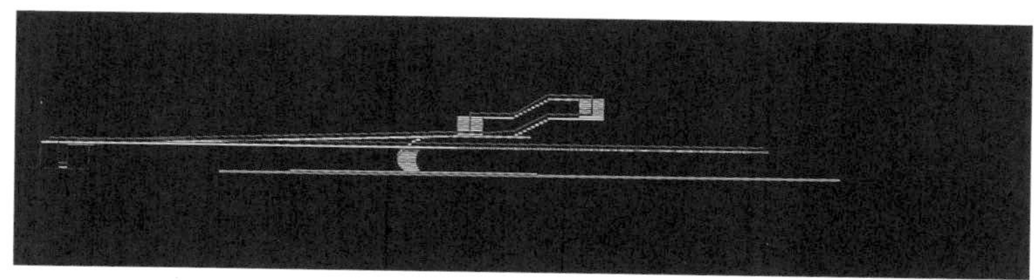

图96
左立面

Fig. 96
Left view

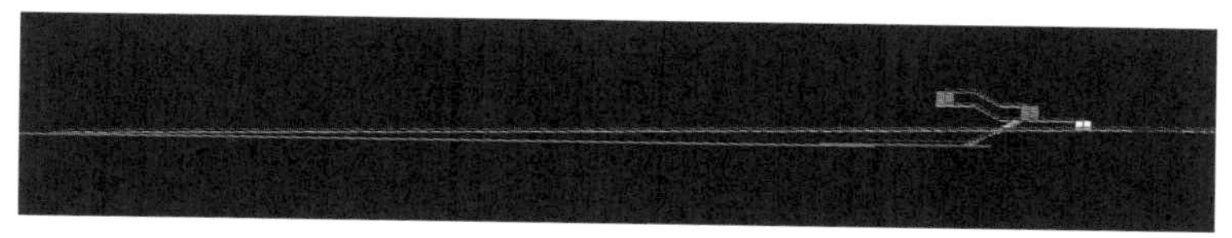

图97
正立面

Fig. 97
Front view

最高点，每个人经历的这些生动如画的空间，都可以在脑海中勾画出完整的建筑面貌。

看过建筑周边环境的全景以及建筑之后，人们将再次沿着狭窄的台阶下楼，并在下一个生动如画的瞬间到来之前，再一次经历幽暗的“视觉静谧”。这种设计是为了突出重要的时刻——生动如画的瞬间。类似的手法可以在作曲家亨德尔的《弥赛亚》结尾处的合唱阿门中体会到，在高潮到来之前加入一秒的静谧，从而更加烘托出高潮的效果。

正如安藤在混凝土围墙旁设计了走道，让人可以沿着围墙往返；台阶的设计也可以双向回转，即上行的台阶逆时针回转，下行的台阶顺时针旋转。

如此复杂的路线设计，安藤旨在让人们失去方向感，从而使得人们产生空间

a departure point, and the second was the point after passing through the free-standing wall. They know the distance they walked and the directions because they walked the same distance twice in opposite directions by walking along the rear side of the freestanding wall and then the opposite side of it.

All this complicated sequence is a device designed to make the visitors can experience the site in various layers. At the highest spot, they are able to organize all the picturesque information obtained from the experience and complete the totality of this building space in their minds.

Then, they walk down the narrow staircase. After taking a look at the panoramic landscape and buildings, they have another blank moment in the dark before the next picturesque moment. It is a way to emphasize the important moment – what Handel composed one second of silent moment before the climax in the "Amen

错觉，认为体验到的空间比实际的更加阔绰。沿楼梯而下的人们将进入最后一个如画的场景——礼拜堂。从礼拜堂内望出去，视野的中心是十字架，它的背景为自然风景，在这之间再无任何阻隔。这与走过围墙时所看到的一字排开的水池、十字架、礼拜堂的静止感受大不相同，因倒映在水面上的十字架的影像，感觉比实际十字架更近。经过安藤设计的这一组空间历程，周边的自然、十字架和建筑的关系持续地变化着。最后，自然、十字架和建筑彼此重叠，共同实现一个如画般的景致，而这三者之间的关系才是水之教堂的设计初衷。

Chorus" in the *Messiah* – to maximize the effect of the climax.

As Ando set up a passage along the freestanding wall to make people walk on both directions, he designed the staircase so that it rotated in both directions. The ascending staircase rotates counter-clockwise, and the descending staircase rotates clockwise.

Ando's intention is to make the visitors lose their sense of direction and feel the space as larger than what it really is. After walking down the stairs, the visitors reach the final main picturesque point, the chapel. When looking outside from inside the chapel, landscape becomes a background and the cross is in the center of their view. There is no obstacle between them and the cross. The relationship among them is different from the contemplative relationship among the water, cross, and chapel which looked like standing in a row when the visitors passed

安藤说："建筑是一种媒介，它让人能够感受到自然的存在。庭园中，大自然每天都在变化，庭园是建筑袒露出的生命核心，传递出诸如光、风和雨等自然现象。"[6]安藤的这一建筑理念在水之教堂的设计中体现得淋漓尽致。水之教堂不仅仅是一幢建筑，它使人们能够像摄像机一样，观察到并记录下空间中每时每刻都在不断变化的、风景如画的瞬间。

水之教堂中令人难忘的瞬间展现出了自然与建筑之间的关系。这种瞬间是借助人在建筑空间中的体验实现的，如沿着混凝土围墙前行，或是顺着台阶上下时。沿着混凝土围墙前行能够借助水平的移动获得不同视点的景观，而上下台阶则通过垂直的移动获得同一地点不同高度的景观。人们借助计算走过的楼梯台阶或脚步的多少感知空间与景观的变化。

the wall. Due to the reflection on the water, the cross looks even closer. During the journey, the relationship among nature, the cross, and architecture changed continuously. At the last picturesque moment, however, the nature, the cross, and architecture are overlapped to accomplish the relationship which is proper to the original purpose of the chapel.

Ando said, "*Architecture is a medium that enables man to sense the presence of nature. And in the courtyard, nature presents a different aspect of itself each day. The courtyard is the nucleus of life that unfolds within the house and is a device to introduce natural phenomena such as light, wind and rain.*" [6] This philosophy is clearly seen in *the Chapel on the Water*. It is not just a building but a device to make visitors keep catching the picturesque views like a movie camera. The ele-

---

[6] **Tadao Ando. "Mutual Independence, Mutual Interpenetration."** ***Nihon no Kenchikuka*****: Vol. 6. (1986).**

安藤的设计手法与汉弗莱·雷普顿测量变量的测绘技术十分相似。测绘师收集垂直和水平方向上的数据，在绘制地图时将各种数据进行综合。在水之教堂，参观者们则是在自己的脑海中完成地图。

ments within each scene show the relationship that changes moment-by-moment.

There are several picturesque moments to capture the relationship between nature and architecture at t*he Chapel on the Water*. These picturesque moments are related to the architectural experiences such as walking along the wall, ascending and descending stairs. Walking along the freestanding wall is a horizontal movement to gather picturesque data from different locations. Ascending and descending stairs make vertical changes of the survey point. Visitors perceive these changes through their bodies, by counting the number of steps or footprints, for example.

With a mere difference in the way of measuring the extent of change, it is similar to Humphry Repton's survey technique. Surveyors collect data containing vertical and horizontal changes and combine all the data when they draw a map. At the *Chapel on the Water*, the map-drawing process is happening in visitors' minds.

风之教堂（1985—1986年）

日本的传统建筑重视入口通道的设计，总是设计得非常独特，最大化地提升人们看到建筑主体前的期待感，并增加紧张感。

例如，举行茶道的茶室总是位于建筑最隐秘的位置，来客要走过一条长长的通道（Roji，日语“路地”），穿过庭园才能到达，而且需要穿过很多道门。相似的空间序列在风之教堂中得以再现。

人们自建筑的北侧，沿着对角线方向进入风之教堂。正如在水之教堂中阻挡视线的混凝土墙一样，入口通道右侧的视野被酒店建筑遮挡，只有左边可以看到自然景观。继续前行，构造复杂、由玻璃廊和混凝土盒子组成的风之教堂才逐渐展露在眼前。

2. Chapel on Mt. Rokko (1985-86)

In traditional Japanese architecture, there used to be an approach path specially designed to maximize the expectation and raise tension of visitors before they experience the main building.

For example, the visitors are supposed to pass the garden which is known as "Roji" to get to the pavilion, in which the tea ceremony is held, in the innermost corner of the building. Passing through the garden, the visitors enter several gates before finally reaching the pavilion. A similar spatial sequence is seen in the *Chapel on Mt. Rokko*.

The visitors approach the *Chapel on Mt. Rokko* following a diagonal path from the north. As the view was hidden by the concrete wall in the *Chapel on the Water*, the right view is blocked by a hotel and the landscape is seen only on the left.

图98
安藤忠雄设计的风之教堂
（兵库县，神户市，六甲山，1985—1986年）

Fig. 98
Tadao Ando. *Chapel on Mt. Rokko,*
1985-1986, Kobe

走到下行台阶的终点处，看到的是玻璃围合的管状长廊。一旦踏上向左转90°的第二段台阶，一侧的自然景观便映入了眼帘。再走下180°回转的第三段台阶，则可以看到另一侧的景象。

至此，人们分别从三个不同的高度和方向领略了教堂周边的自然景色，这个渐进的序列是进入教堂空间的序曲。这一入口通道的空间组织与京都17世纪建成的日本皇室夏季的行宫——桂离宫的入口组织非常相似。

建于江户时代的桂离宫，要经过“路地”庭院才能进入。入口通道被分为曲折的几段，人们得以在经过精心架构的每一个位置与角度上都能观赏到周边的景观。这样的布置不仅仅使人们经过空间的转移失去方向感，在心理上获得比实际空间更加广阔的感受，也在人们进入主要建筑空间前，展示出丰富的如画般的景

Gradually, the visitors see the *Chapel on Mt. Rokko* complex composed of a glass corridor and a concrete box-shaped chapel.

When they reach the end of the descending stairs, their view is filled with a glass tube-shaped corridor. But to step on the second staircase, they turn 90 degrees to the left and the surrounding landscape is displayed in front of them. After that moment, they turn 180 degrees to make a U-turn and get to see the other side of the landscape.

So far, they have three different views of the surrounding landscape at three different heights and axes. This approaching sequence is a prelude before entering the main building. This kind of approach was also used at the *Katsura Villa*, which was built in the 17th century in Kyoto as a summer house of Japanese emperors. Built during the Edo period, the Villa was approached through the Roji garden.

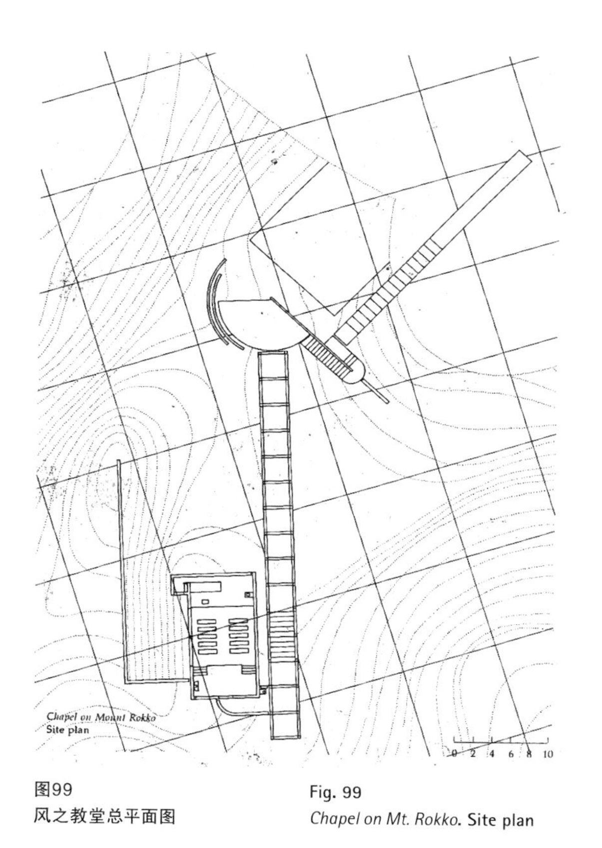

图99
风之教堂总平面图

Fig. 99
Chapel on Mt. Rokko. Site plan

观效果。复杂曲折的路线甚至可以阻挠擅闯者的行动。

安藤希望人们在进入风之教堂前，借助于方向的两次转折（台阶阶段）和高度的三次变化，欣赏到建筑周边的景观。当人们拾级而下，来到一个比常规休息平台略大却够不上广场的空间。在这里，还需要左转135°才能进入风之教堂的长廊，这是人们真正进入建筑之前的第三个转向点。转身135°后，人们会再次为所看到的周边景色感叹，然后走进玻璃围合的长廊。走在长廊里，尽头的景观如同画框内一点透视的油画，此地的空间感与约瑟夫·阿尔伯斯（1888—1976）的《方形礼赞》（1964年）十分相似。

入口通道用的是半透明的玻璃，营造出漫射光的效果，与日本传统建筑中自然光透过和纸的效果极为相似。进入玻璃廊道前，在人们的视野中，自然景观是背

Since the approaching path was fragmented and angled, the visitors can enjoy various surroundings specially framed. It is not only an intention to make visitors disoriented and feel that the space is larger, but also a device to present them with different picturesque views before getting into the main building. The complicated movement path would also even confuse intruders.

Ando's intention is to allow visitors to enjoy the surrounding view at various heights and directions by experiencing three different heights and two turns before entering the main building. When they stepped down the stairs, they arrive at a small space which is rather big for a landing place and small for a plaza. At this point, they must turn 135 degrees to the left to enter the corridor. This is the third turning point before the actual entrance. Making a 135 degree turn, the visitors get to appreciate the surrounding view once again, and then enter the glass corridor. After entering the corridor, they see the landscape, which looks like a

图100
入口通道
俯视图

Fig. 100
Top view of the approach path

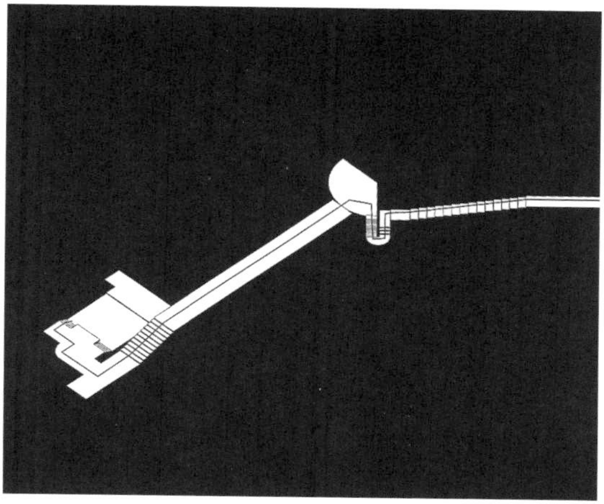

图101
轴测图

Fig. 101
Axonometric view

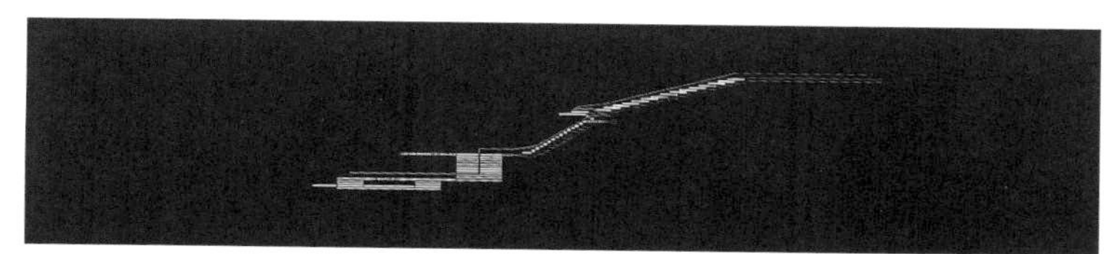

图102
正立面

Fig. 102
Front view

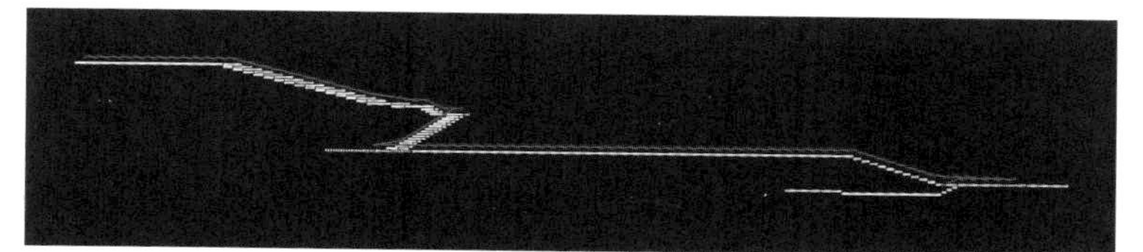

图103
左立面

Fig. 103
Left view

景，建筑位于视线的中心；但是走进玻璃廊道，构图关系发生了逆转，自然景观位于视线的中心，建筑在透过半透明的玻璃射入的漫射光下变得模糊不清。

沿着通道前行，尽端那幅自然景色的"油画"逐渐变大，"画框"从视野中一个个地消失，好像是一点点逐步发现自然的过程。在到达最后一个画框之前，走廊向右转，引导人们进入一个由清水混凝土建成的狭窄幽暗的空间。这个空间与水之教堂逐渐下沉的台阶相似，在两个风景如画的片刻起着过渡的作用。在狭小的空间中，人们被迫向右转90°，在左右两侧混凝土墙的"夹击"下，只能正对着前方的清水混凝土墙前行。

片刻之后，人们终于来到了长方形的礼拜堂，还可以透过窗口看到礼拜堂左侧的一抹绿色——庭园。通过十字架形状的窗口可以看到细心修剪过的庭园——近

painting in a frame, at a vanishing point. This kind of feeling of space is just like Josef Albers's *Homage to the Square* (1964).

This glass corridor is constructed with translucent glass to create a diffused lighting effect, which is similar to the light coming through the window covered with *shoji* paper in traditional Japanese architecture. Before getting to the corridor, the landscape was a background and the building was in the center of the visitors' line of vision. But once entering the corridor, the nature is in the center of the line of vision and the building looks blurred out due to the diffused light coming through the translucent glass.

As walking along the hallway, the framed nature seems to get bigger, and the frames disappear one by one. It is a process of finding nature little by little. Just before passing the last frame, visitors have to turn to the right to enter a dark and

处是草坪和两棵树；后面是一面矮墙，看不到地面；矮墙上露出远处作为背景的几棵大树。

从入口通道至此，自然与建筑之间的关系不断变化着：开始是自然环抱着建筑，接着是建筑的长廊包含自然，最后是自然与建筑隔着窗户交织在一起。

在走进建筑之前，人们从高处俯瞰自然景观与建筑。随后，经过台阶走到稍低的位置，走进玻璃走廊之前，再以全景将各种元素尽收眼底。从左侧开始展开的是自然景观、框架中的景观，然后是建筑、庭园和矮墙，最后又是自然。伴随着人们一步步走入建筑，建筑与自然的关系不断地转换着。从这一角度说，教堂成为自然与建筑持续转换的媒介。

根据冈特·尼契克“时间就是金钱，空间就是金钱”的理论，像美国这样拥有

narrow space constructed with bare concrete material. This is a filtering sequence between two picturesque moments like the descending staircase at the *Chapel on the Water*. The visitors are supposed to turn 90 degrees to the right in this small and narrow space. Blocked by left and right walls, they are driven to the front which shows only concrete walls.

After the moment, finally, they reach large rectangular space of the chapel building with a garden view on its left. Through the window, of which the frame is cross-shaped, they see an artificially trimmed garden. The foreground is composed of lawn and a couple of trees, and the middle ground is hidden by a freestanding wall. In the background, some trees can be seen over the wall.

During the journey, the relationship between the building and nature changes: The nature embraces the building at first, and then the building includes the na-

图104
风之教堂景观序列

Fig. 104
Picturesque sequence of the ***Chapel on Mt. Rokko***

富足空间的地区，由于时间更为重要，所以建筑开发的重点就是缩短时间。以高速公路为例，它就是为了缩短在遥远的城市之间旅行的时间而建造的。

尼契克认为，像日本这样的岛国，没有充足的空间，时间反而富裕些。因此在那样的地区，建筑设计的重点就集中在延迟时间以拓展空间上。就像日本传统的庭园设计，将空间分隔、转折、迂回，以延长在其中行走的时间，让人们感受到比实际空间更为广阔的空间。

我完全认同尼契克关于日本传统建筑空间"消磨时间"的理论与分析方法。日本传统建筑中设计复杂的入口通道——或多段组合，或曲折，自然环境与建筑之间的交替主要是通过水平方向的运动实现的。利用台阶实现的竖向运动的主要目的是为了适应地形的高低变化。

ture in the corridor, and finally the nature and the building are interwoven in layers.

Before entering the building, visitors look down on the landscape and architecture from above the building. They see all the elements again before entering the glass corridor. From the left, the nature, framed nature, chapel building, garden, free-standing wall and nature are displayed before their eyes. When they walk through the building, the relationship between nature and building keeps shifting. In this regard, a building is a medium to alternate the relationship between nature and architecture.

According to Gunter Nitschke's *Time is Money – Space is Money,* architecture has developed focusing on saving time in a vast area like the United States because it is time that is important in that region. As an example, a highway system was

然而，在安藤的建筑中，竖向的垂直变化表现得更为积极。这可以理解为安藤在日本传统建筑“消磨时间”的二维平面的手法上增加了三维的竖向变化。从水之教堂的设计中可以看到，安藤嵌入的上行与下行的台阶不只是为了顺应地形，而是为了“消磨时间”、引导景观序列，如此便可从不同的高度演绎出变幻万千的如画风景。

测量大地高程时，测绘工程师在某一点测量出基准高度，然后或上移或下降一个高度到第二个点，第三个点……再根据得到的一系列高度变化的数据，描绘出客观的地形。安藤似乎将访客的眼睛想象为测量仪。对于他来说，人们的参观路径与测绘师测量地形的过程是相同的。

在安藤的建筑中，台阶不再是传统的仅仅为适应地形变化而设置的台阶。安藤

developed to cut travel time between distant cities.

Nitschke insists that there is insufficient space and plenty of time in islands such as Japan, and the architecture in that region has accordingly developed focusing on delaying time to maximize the space. For example, in Japanese traditional garden design, there are devices to make the space fragmented, angled, and detoured to delay the time to walk in the garden so that people perceive the space as larger than real.

I totally agree with Nitschke's "killing time" theory and the method for it. The approaching path was made complicated in Japanese traditional architecture, and the change in the relationship between nature and buildings is due to the horizontal movement. The approaching path is fragmented and twisted. The vertical changes such as stairs were made only to accommodate the topographical changes.

的台阶不仅仅引导竖向的垂直运动，同时也引导水平的运动。一般台阶上每个踏步的高度与宽度是固定不变的，而阶段就是多个踏步的重复。人们通过身体感知水平或垂直的运动，以及在台阶上不同的高度看到的不同场面，获得更为客观的空间感知。

安藤忠雄设计建筑目的并不在于建造一个实体，而在于建造一个自然与建筑共同组成的立体构成。为了让人们体验他的建筑空间，安藤在建筑内外建造了一条复杂的路径。这条路径不仅在水平方向上分隔、组合、回转，还以台阶连接不同高度的路径，构成垂直方向的变化。沿着如此复杂的路径，人们可以体验到自然与建筑之间不断变换的关系，以及变幻的景致。客人们经历着如画的场景变幻，头脑中构筑着完整的空间。在安藤的建筑中，客人的身体成为空间的测量仪，而建筑则成

In Ando's design, however, the vertical changes appear more actively. Ando adds three-dimensional vertical changes into the Japanese two-dimensional "killing time" method. In the *Chapel on the Water*, he inserted ascending and descending stairs – not for a topological reason but for a delayed and panoramic sequence. By doing so, he could create various picturesque scenes from different heights.
Using a height survey technique, engineers measure a reference height from one point, and then ascend or descend to the second point to measure the difference of the height. The accumulated data create a more objective topology of the land. Ando assumes a visitor's eyes as a survey device and the visitor's movement path as a movement path in a survey process.

Stairs in Ando's buildings are not just stairs. Stairs are designed not only for a vertical movement but also for a horizontal movement. Since the stair risers and treads of the same size are repeated, people feel the horizontal and vertical

为这种移动的测量装置活动的环境。

沿着安藤建筑中的路径，人们体验着建筑元素与自然景观之间一系列的持续变化。这种重视关联价值的设计看似来自安藤的东方文化背景。他的建筑通过围墙或长廊与自然交融，正如围棋中一方的棋子落下使其领域得到扩张。安藤通过围墙等建筑元素使建筑与自然景观相融合。但是与传统的东方建筑流动的空间不同，安藤的建筑空间由墙体分隔围合，具有明确、肯定的边界。例如，风之教堂就是由两个盒子空间组成的：一个是窄长的玻璃走廊，另一个是混凝土盒子——礼拜堂。

安藤的建筑与其说是以柱网结构为主，不如说是以承重墙体系为主的，营造出传统的西方建筑空间。在教堂前的庭院中，低矮的围墙与远处的借景表明了清

movements through their bodies and get to have more objective feelings of space with different scenes they see before and after the stairs.

The aim of Tadao Ando's architecture is not to create an object, but to make a three-dimensional composition of nature and architecture. Ando sets up a complicated path inside and outside the building to have visitors explore the space. The horizontally fragmented and twisted path is connected with stairs which make vertical changes. Following the complex path, the visitors are able to experience various views showing the different relationships between nature and architecture. Experiencing the picturesque sequences through their bodies, they accomplish the totality of the space in their minds. In Ando's architecture, a visitor's body becomes a survey device and architecture becomes a medium for moving the body of device.

晰的东方风格的空间。安藤的建筑空间虽然具有很多十分传统的西方建筑空间的特征，同时也能发现，其平面布局是非对称的自由平面，内含静谧的、枯山水风格的虚空空间。安藤忠雄是最成功地将东西方文化杂合出现代建筑的建筑师：他使用20世纪最典型的建筑材料——混凝土，他创造的建筑空间具有东方建筑特征——积极、自由地与大自然交流，也拥有西方建筑特征——具有几何特征的平面与剖面。

Walking along the path of Ando's buildings, people experience the sequence of continuously changing relationships between architectural elements and landscape. The emphasis on relationships seems to come from his Eastern cultural background. His buildings empathize with nature through freestanding walls or long hallways. As the territory expands by placing stones in Go, his buildings merge with nature through architectural elements such as freestanding walls. Unlike the fluid, void space of traditional Eastern architecture, however, Ando's architectural space is clearly divided by walls. For example, the *Chapel on Mt. Rokko* is composed of two main box spaces; one is a long and narrow glass corridor and the other is a concrete chapel.

Ando's architecture is more based on a bearing wall structure than on a column structure, creating Western style void space. But the garden in front of the main chapel is clearly Eastern style void space framed by freestanding walls and the landscape beyond the low wall. His architectural space shows a great deal of traditional Western style void space, and at the same time, his plans are non-symmetrical free plans with Zen-type silent void space. Tadao Ando is an architect who created the modern hybrid of Eastern and Western culture most successfully: Using concrete, which represents the 20th century, he created architectural space that represents Eastern architecture that actively and freely communicates with nature, and Western architecture that has geometrically defined plans and sections all together.

结语

Conclusion

建筑涉及空间，而空间由虚空构成。通过建筑历史我们了解到，每种文化对空间都有独特的理解、阐释和设计。不同文化的哲学能在其建筑空间中得到反映。当两种文化产生交流时，建筑空间也会自然地发生变化。

人们大多把20世纪发端于欧洲的现代主义看做是工业革命和功能主义形成的思潮。本书摒弃了从技术出发的观点，从东西方文化交流的角度重新审视现代主义。20世纪的现代主义是从16世纪起远东亚洲文化大规模输入西方的过程中发生的文化杂合的产物。

东方和西方基于互不相同的哲学，走过各自的发展路程。有趣的是，建立两种文化哲学根基的哲学家们都在公元前570年至公元前300年之间诞生。代表西方的哲学家有：毕达格拉斯、柏拉图、欧几里得，东方则有老子、孔子、释迦

Architecture is about space which is made of void. We have seen that the different perceptions and interpretations of space in different cultures bring about different design throughout the history of architecture. The different philosophy of each culture is reflected in its architectural space. When there is cultural exchange between two cultures, their architectural spaces naturally change, too.

Modernism that appeared in the 20th century in Europe is usually interpreted as a trend emerging from the Industrial Revolution and functionalism. In this book, modernism was studied with regard to the cultural exchange between the East and West rather than from a technological point of view. Modernism of the 20th century is a cultural hybrid that appeared when Far East Asian culture was introduced to Europe on a large scale from the 16th century.

The East and the West have developed based on different philosophy. Coinciden-

牟尼。在西方，柏拉图的理念学说和毕达格拉斯的数学思维支配了整个西方思想，认为数学是抵达完美的理想——理念的道路，这种认识成为了西方文化的根基。在宗教建筑等象征权力的建筑中，几何学和数学成为构成建筑空间不可或缺的因素。罗马万神庙和圣索菲亚大教堂的平面和剖面是这种趋势的代表。总之，在20世纪以前修建的西方所有重要建筑，其空间都可用几何学和数学来分析。

图105
本书意象图

Fig. 105
Concept collage image of this book

tally, the philosophers who laid the foundation for the two cultures were born between 570 B.C. and 300 B.C. In the West, there were Pythagoras, Plato, and Euclid and in the East there were Lao Tzu, Confucius, and Buddha. Plato's Idea ideology and Pythagoras' mathematical ideology dominated Western ideology, and the thought that there is a perfect ideal of Idea which can be reached through mathematics became a basic thought of Western culture. Mathematics and Euclid's geometry became an essential element in the composition of space in architectural buildings that represent the symbolism of power, such as in religious architecture. The plans and sections of the *Pantheon* in Rome and *Hagia Sophia* are examples of this tendency. All such dominant architectural buildings built before the 20th century in the West can be analyzed geometrically and mathematically.

Unlike the West that generally considers the void as negative; the East considers that void has a supreme value and the values of everything lie in relationships.

在西方一般被解释为否定意义的“空”，在东方却被视作是最高价值，所有事物的价值都系于“关系”。说到“空间”，英语单词“space”也包含遵循数学规则的宇宙的意思，东方文化则指的是“虚空之间”，即空间的意义在于“空”和“关系”。以“关系”为中心的思维方式，还反映在中国汉字的构成原理中。“木”和“一”这两个字的组合，可以根据笔画位置的不同，组合成“本”、“末”等不同汉字。而英语单词的构成，是把字母沿着一条线排列，更换字母顺序，组成新的单词。再举其他的例子——在两种文化中最受欢迎的游戏——围棋和国际象棋。围棋的黑色和白色棋子，根据相互位置关系决定其地位，被叫做“目”的“空的地方”多的一方胜。国际象棋则先定好每个棋子的地位，这些棋子按照几何形的轨道移动，先吃掉对方棋子的一方获胜。基于不同的文化背景，东方和西

When expressing "space," whereas the word "space" in the West also means cosmos, which implies having mathematical rules, it means "a gap between voids," focusing on "void" and "relationship" in the Eastern culture. The relationship-oriented thought is also seen in the composition principle of Chinese characters; for example, the characters "木" and "一" create different words by changing their positions. In the composition of English words, the alphabetical letters are arranged along an axis and new words are formed by changing their order. In the most popular games in each culture, in the Go game in the East, while the ranks are decided by the mutual positioning relationship between white stones and black stones and the player who has more "void," which is called as "house," wins – in the West in the game of chess, the power ranks are already decided and the player moves along the geometric path and wins the game by killing the opponent's king. Based on these kinds of cultural differences, the East and the West evolved their own architectures without exchanges.

方各自发展出了自己的建筑。

东西方早在罗马帝国和汉朝就通过丝绸之路进行过文化交流，但全面的文化交流直到16世纪才开始。此后，英国通过瓷器受到中国文化的影响，把瓷器叫做china。在当时，很多学者访问中国，把中国当做自己梦想中的理想国，可见中国文化对西方人的冲击有多大。17世纪末，传教士和商人们翻译出版老子和孔子的书，介绍给欧洲社会，早期停留在表面的以模仿为主的文化交流逐渐成熟，西方人开始了解东方思想的本质。改变最早始于园林设计。在风景式园林的影响下，英国园林开始摆脱几何学、追随中国风格的非几何式设计。中式宝塔和亭子开始经常出现在英国园林中。从风景式园林的创始人汉弗莱·雷普顿的草图中可以看到，园林中的元素会依照亲历者视角位置的变化而改变相互关系。在那个时

There were cultural exchanges between the two cultures via the Silk Road from the Roman Empire and Han Dynasty periods respectively, but full-fledged exchanges only began in the 16th century. Calling ceramics "china", England was influenced by China through tableware. At that time, the cultural influence from China was so powerful that it drew many European scholars to China, who returned to Europe describing China as the ideal land they had dreamed of. Toward the end of the 17th century, when many missionaries and traders translated the books of Lao Tzu and Confucius and introduced them to Europe, the cultural exchange, which until then was merely a superficial imitation, developed to become a better understanding of the essence of Eastern thought. The change started from garden design. Under the influence of the Picturesque Movement, the existing geometric England garden began to adopt non-geometric Chinese-style design, and Chinese-style pagodas and pavilions appeared in lots of gardens. The sketch by Humphry Repton shows the relationship between elements that change

候，“关系”就已经成为了西方建筑师的设计中的重要元素。

众所周知，19世纪末起，东方对西方艺术产生了很大的影响。考尔德在1930年的作品《美杜莎》表现出过去西方雕塑历史上不曾有过的“空间”。《莫比尔》表现的则全是随着各种元素的变化而改变的关系。此外，亨利·摩尔、保罗·克利、蒙德里安等人的作品也清晰地表现出东方文化的价值——“虚空”和“模糊的边界”。

艺术领域发生变化后，赖特、密斯和柯布西耶开始将东方价值融入自己的建筑中。日本浮世绘在美国的主要代理商赖特，明确承认自己的建筑受到日本建筑的影响。密斯收藏中国哲学书籍早已广为人知，通过与赖特的交流，受到东方建筑的影响。尽管他本人没有正式表示过，但在1937年与中国建筑师李承宽的

according to the position of the first-person viewer. By this time, "relationship" became an important element in designs of Western architecture.

The influence of the East on Western art toward the end of the 19th century is well- known. Calder created void space, which till then was not found in the history of Western sculpture, in his *Medusa*, and his famous *Mobil*, is all about the relationship that changes according to the movements of many elements. In addition, Henry Moore, Paul Klee and Mondrian expressed "void" and "ambiguous boundary," which are characteristically Eastern values, in their artworks.

After the changes in art, Frank Lloyd Wright, Mies van der Rohe, and Le Corbusier started to incorporate Eastern values for the first time in their architecture. Wright, a major supplier of Japanese folk painting in the United States, definitely announced that his architecture was influenced by Japanese architecture. Mies

对话中，密斯承认自己受到中国文化的影响。密斯的早期作品乡间砖宅与杜斯堡的《俄罗斯舞蹈的韵律》气质类似。如果将密斯的代表作巴塞罗那博览会德国馆的照片与日本传统建筑的照片并排放在一起比较，很容易看出令人不可思议的相似之处。

柯布西耶的作品更明显地显示出，建筑空间从西方价值转向东方的特点。1922年的早期作品沃卡松别墅，是可以用几何学分析的传统西方建筑平面。到了萨伏伊别墅，柯布西耶开始摆脱了几何学的平面。此后的米尔住宅的平面则内外空间界限模糊。其后期作品卡彭特中心的正立面中心留有空洞，将建筑空间当作了第一人称经验的产物来设计。

欧洲的现代主义是西方世界从20世纪初开始接受东方思想，进而发展出新

collected Chinese philosophy books, and he became influenced by Eastern architecture through his exchanges with Wright. Though he did not officially announce it, he admitted that he was influenced by Chinese culture during his conversation with Chinese architect Chen-Kuen Lee in 1937. One of his early works *Brick Country House* resembles *Rhythm of Russian Dance* by Theo van Doesburg, and the picture of his masterpiece *Barcelona Pavilion* looks incredibly similar to that of the Japanese traditional architecture.

Le Corbusier displays a clearer transition of architectural space from Western values to Eastern values. The plan of one of his early works *Villa Vaucresson*, designed in 1922, is a traditional Western architectural plan that can be geometrically analyzed. From *Villa Savoye*, however, he broke from geometric plans, and used a plan with a ambiguous division between the inside and outside in *Mill Owner's House*. The design of his later work *Carpenter Center* shows that the middle of the front elevation of the

的建筑。那么在东方到了80年代，才有安藤忠雄这样的建筑家设计出受到完整西方思想影响的建筑。安藤忠雄的建筑，可以说是以“关系”为中心的东方价值观和以几何性的西方价值观相结合的典范。

建筑不能简单地从经济或技术的角度理解，建筑包容了人所有的活动。它是每个时代的文化的结晶，是人类生活的产物。因此，必须从文化的视角阅读建筑的历史。本书试图从文化的角度解读20世纪的现代主义建筑运动，而非技术性的工业革命。

大自然中为什么存在雌雄两性？生物学家们解释说，是为了组合互不相同的遗传因子，将优良基因遗传给下一代而自然形成的结果。在不同地区和环境中生长的雌雄两性，在交尾季节生产出更优良的下一代，这就是大自然的规律。

building is void and architectural space is made by subjective experience from a first-person point of view.

While modernism in Europe created new architecture by adopting Eastern thought since the beginning of the 20th century, the architecture that shows the well-established influence of Western thought appeared in the East in the 1980s, through architects such as Tadao Ando. The architecture of Tadao Ando is an example that harmonized relationship-oriented Eastern values and geometric Western values.

Architecture cannot be understood by simple perspective of economy and technology. Architecture contains all the aspects of human behavior. It is crystallization of each era's culture; byproduct of human life. Therefore architectural history must be read through a cultural viewpoint. This book is a try to interpret Modern-

同样的过程也发生在文化领域。两种不同的文化，数千年来按照各自不同的思维，发展出了互不相同的建筑空间。到了20世纪，东方的文化基因流向了欧洲建筑，衍生出新的建筑文化——现代主义建筑。20世纪的多位建筑大师很好地将东方和西方文化融合起来，结出现代主义建筑的果实。那么，接下来，建筑进化的下一阶段又将是什么呢？

ism of 20th century movement through a cultural viewpoint instead of that of technological Industrial Revolution.

Biologists claim that the reason why there are two sexes in nature is to hand down superior genes to the next generations by combining different genes. After growing up in different areas and environments, two sexes meet and are united during the breeding season to produce better offspring.

The same process happens in cultures, too. The Eastern and Western cultural characteristics were respectively formed and evolved in a separate region. In the 20th century, the cultural gene of the East flew into European architecture to create a new architectural cultural hybrid, which is modern architecture. Several great architectural masters delivered successful results by splendidly combining the Eastern and Western cultures. What, then, will be the next step in the world's architectural evolution?

参考书目 Bibliography

建筑 Architecture

Alessandra Latour. *Louis I. Kahn, Writings, Lectures, Interviews.* New York: Rizzoli, 1991.
Amos Ih Tiao Chang. *The Tao of Architecture.* Princeton: Princeton University Press, 1956.
Anita Berrizbeitia & Linda Pollak. *Inside Outside.* Gloucester: Rockport Publisher, 1999.
Colin Rowe. *The Mathematics of the Ideal Villa and Other Essays.* Cambridge: MIT Press, 1999.
David B. Brownlee & David G. De Long. *Luis I. Kahn: In the Realm of Architecture.* New York: Rizzoli, 1992.
Francesco Dal Co. *Tadao Ando Complete Works.* London: Phaidon Press, 1996.
Harold Alan Meek. *Guarino Guarini and His Architecture.* New Haven: Yale University Press.
Ignasi de Sola-Morales. *Mies van der Rohe Barcelona Pavilion.* Barcelona: S. A., 1993.
John Lobell. *Between Silence and Light.* Boston: Shambhala, 2000.
Kevin Nute. *Frank Lloyd Wright and Japan.* London: Routledge, 1993.
Lluis Casals. *Reflxions Mies.* Barcelona: Triangle Postals, 1999.
Peter Eisenman. *"miMises READING: does not mean A THING"*, in *Mies Reconsidered.* Chicago: The Art Institute of Chicago, 1986, pp. 86-98.
Philip Drew. *Church on the Water Church of the Light.* London: Phaidon Press, 1996.
Urs Buttiker. *Louis I. Kahn Light and Space.* Basel: Birkhauser, 1994.
Werner Blaser. *Mies van der Rohe Fansworth House.* Basel: Birkhauser, 1999.
Werner Blaser. *West Meets East Mies Van der Rohe.* Basel: Birkhauser, 2001.
Willy Boesiger. *Le Corbusier.* Basel: Birkhauser, 1972.
Willy Boesiger. *Le Corbusier 1910-65.* Basel: Birkhauser, 1972.
Yoshinobu Ashihara. *Exterior Design in Architecture.* New York: Van Nonstrand Reinhold Company, 1970.

建筑历史 Architectural History

Bates Lowry. *Renaissance Architecture.* London: Prentice-Hall International, 1962.
Bill Risebero. *The Story of Western Architecture.* Cambridge: MIT Press, 1985.
Eleanor von Edberg. *Chinese Influence on European Garden Structures.* New York: Hacker Art Book, 1985.
Elisabeth B. Macdougall. *The French Formal Garden.* Washington: Dumbarton Oaks, 1974.
Hillary French, *Architecture.* New York: Watson-Guptill Publications, 1998.
Humphry Repton, *Observations on the Theory and Practice of Landscape Gardening.* Oxford: Phidon, 1980.
Karin Kirsch. *The Weissenhofsiedlung.* New York: Rizzoli, 1989.
Lawrence Wodehouse & Marian Moffett. *A History of Western Architecture.* Mountain View: Mayfield Publishing Company, 1989.
Mavis Batey. *Alexander Pope.* London: Barn Elms Publishing, 1999.
Nikolaus Pervsner. *The Picturesque Garden and Its Influence.* Washington: Dumbarton Oaks, 1974.
Osvald Siren. *China and Garden of Europe.* Washington: Dumbarton Oaks, 1990.
Rudolf Wittkower. *Gothic vs. Classic, Architectural Projects in Seventeenth Century Italy.* New York: George Braziller, 1974.
William L. MacDonald. *The Pantheon.* Cambridge: Harvard University Press, 1976.

东亚建筑 East Asian Architecture

Gunter Nitschke. *From Shinto to Ando*. London: Academy Group, 1993.
Gunter Nitschke. *Japanese Gardens*. London: Taschen, 1999.
Julia Meech. *Frank Lloyd Wright and the Art of Japan*. New York: Harry N. Abrams, 2001.
Kazuo Nishi. *What is Japanese Architecture?*. Tokyo: Kodansha, 1983.
Kevin Nute. *Frank Lloyd Wright and Japan*. London: Routledge, 2000.
Mitchell Bring. *Japanese Gardens*. New York: McGraw-Hill Company, 1981.
Norman F. Carver Jr.. *Form and Space of Japanese Architecture*. Tokyo: Shokokusha, 1955.
Sutemi Horiguchi. *The Katsura Imperial Villa*. Tokyo: The Mainichi Newspapers, 1953.
Teiji Ito. *The Japanese Garden*. New Haven: Yale University Press, 1972.

哲学 Philosophy

金容沃. 这就是老子哲学. 首尔: Tongnamoo, 1989.
佚名. 怎样研究东方学. 首尔: Tongnamoo, 1990.
裴约翰. 西方思想史. 首尔: 同人书院, 1996.
Ju,Chilseong. 东亚传统哲学. 首尔: 艺文书院, 1998.
Fritjof Capra. *The Tao of Physics*. Boston: Shambhara, 1975.
Ren Jiyu (任继愈) . *A Taoist Classic: The Book of Laozi*. Beijing: Foreign Language Press, 1993.

历史 History

Adolf Reichwein. *China and Europe*. London: Kegan Paul, Trench, Trunbner & Co., 1925.
David E. Mungello. *The Greatest Encounter of China and the West, 1500-1800*. Lanham: Rowman & Littlefield Publishers, 1999.
Liddell MacGreagor Mather. *The Key of Solomon The King*. The Book Tree, 1999.

艺术 Art

Dawn Jacobson. *Chinoiserie*. London: Phaidon Press, 1993.
The Solomon R. Guggenheim Museum. *Paul Kee at the Guggenheim Museum*.

图书在版编目（CIP）数据

现代主义：东西方文化的杂合 /（韩）俞炫准著；
（韩）太贞姬，江岱 中译；（韩）崔胤雅，（韩）俞炫准 英
译. —— 上海：同济大学出版社，2012.7
（倒影）
书名原文：Modernism: A Hybrid of Eastern and
Western Culture
ISBN 978-7-5608-4790-0

Ⅰ. ①现… Ⅱ. ①俞… ②太… ③江… ④崔… ⑤俞…
Ⅲ. ①现代主义－建筑流派－对比研究－东方国家、西
方国家 Ⅳ. ①TU-86

中国版本图书馆CIP数据核字(2012)第021637号

现代主义：东西方文化的杂合
（韩）俞炫准 著

责任编辑 江 岱
校 对 徐春莲
设 计 厉致谦

出版发行 同济大学出版社 www.tongjipress.com.cn
地址：上海四平路1239号 邮编：200092 电话：021-65985622
经 销 全国各地新华书店
印 刷 上海雅昌彩色印刷有限公司
开 本 889mm×1194mm 1/32
印 张 6
字 数 161 000
版 次 2012年7月第1次出版 2012年7月第1次印刷
书 号 ISBN 978-7-5608-4790-0
定 价 36.00元